Status of Recent Geoscience Graduates 2016

Carolyn Wilson
American Geosciences Institute
Alexandria, VA 22302

Status of Recent Geoscience Graduates 2016
Edited by Carolyn Wilson

ISBN: 978-0-913312-71-1

Graphs by Carolyn Wilson, AGI Workforce Program
Design by Brenna Tobler, AGI Graphic Designer

For more information on the American Geosciences Institute and its publications check us out at www.americangeosciences.org/pubs.

Carolyn Wilson, Geoscience Workforce Data Analyst
American Geosciences Institute
4220 King Street, Alexandria, VA 22302

www.americangeosciences.org
cwilson@americangeosciences.org
(703) 379-2480, ext. 632

Front cover photo ©Victoria Benson. Field titration of glacial meltwater on Mount Rainier, Washington, to test atmospheric influence of glacial melt.

All photos in this report were submitted to the 2016 Life in the Field contest, which requested images representing meaningful geoscience work through internships, research, employment, or field experiences.

AGI american geosciences institute
connecting earth, science, and people

GEOSCIENCE STUDENT EXIT SURVEY

About AGI's Geoscience Student Exit Survey

The American Geosciences Institute (AGI) launched the Geoscience Student Exit Survey to help geoscience departments assess the educational experiences of graduating students, as well as for AGI to examine the trends of strengths and weaknesses of new graduates entering the workforce. With this survey, we can identify student decision points for entering and persisting in a geoscience field, measure participation in co-curricular activities that support the development of critical geoscience skills, identify the geoscience fields of interest, identify the preferred jobs and industries of graduating students, and establish a benchmark for a detailed study of career pathways of early career geoscientists.

This report examines the responses to AGI's Geoscience Student Exit Survey by graduates from the 2015-2016 academic year.

The survey has four major sections: student demographics, educational background, postsecondary education experiences, and post-graduation plans, with specific questions covering community college experiences, quantitative skills, field and research experiences, internships, and details about their immediate plans for graduate school or a new job. The survey was piloted twice in Spring 2011 and Spring 2012. For Spring 2013 graduation, we opened the survey up to all geoscience departments in the United States. Since 2014, the survey was available for graduates at the end of each semester—fall, spring, and summer. AGI also asks their federation of member societies to send the survey out to their student membership, which, in addition to the numerous departments already distributing the survey, reached a larger pool of recent geoscience graduates.

With awareness of the survey growing, AGI has been able to engage collaborations with Canada and the United Kingdom to send versions of the survey to their graduates as well. AGI will continue to work to try and expand the number of countries that distribute the survey in the future in order to work towards a more global understanding of the preparation of geoscience graduates for the workforce.

To encourage participation, departments, member societies, and international organizations that help distribute AGI's Geoscience Student Exit Survey will receive the data in aggregate for their constituency, as long as they have a sufficient number of participating students to ensure individual response privacy.

If you would like more information or would like your department, society, or country to participate in AGI's Geoscience Student Exit Survey, please contact Carolyn Wilson at cwilson@americangeosciences.org.

Acknowledgements

I would like to recognize several organizations and individuals for their support for this project. Thanks to ConocoPhillips for their financial support of the project this year. Thanks also to the American Geophysical Union, the American Institute of Professional Geologists, the Association for the Sciences of Limnology and Oceanography, the Geological Society of America, and the Society of Exploration Geophysicists for distributing the survey to their student membership. I also want to thank the AGI Workforce 2016 Fall Intern, Kelsey Watson, for her hard work cleaning and organizing the data from this survey, as well as her help coding the qualitative responses and creating the map figures in the report. Finally, I would especially like to thank the contacts from each participating department for distributing the survey to their graduating students.

Executive Summary

The American Geosciences Institute's (AGI) Status of Recent Geoscience Graduates 2016 provides an overview of the demographics, activities, and experiences of geoscience students that received their bachelor's, master's, or doctoral degrees during the 2015-2016 academic year. This research draws attention to student preparation in the geosciences, their education and career path decisions, as well as examines many of the questions raised about student transitions into the workforce.

The Status of Recent Geoscience Graduates report was first released in 2013 presenting data from spring 2013 graduates. A new report has been released each year with the data from the graduates from that academic year. This report presents the results for the end user's consideration. Since beginning the data collection from AGI's Geoscience Student Exit Survey, compelling trends have consistently arisen each year related to necessary experiences for developing critical skills for the workforce.

Over the past few years, participation rates in quantitative courses beyond Calculus II have been consistently low at all degree levels, and the availability of these courses to students depends on the type of institution they are attending. For example, students that attended R1 doctoral institutions were more likely to take at least one of these upper level courses compared to students that attended a liberal arts college. It has also raised the question of availability of these quantitative courses to graduate students considering it is possible that the students working on graduate degrees took these courses, like Linear Algebra and Differential Equations, as an undergraduate. Discussions with industry representatives have indicated that a lack of understanding of advanced math can affect sustained employment or advancement depending on the job expectations.

In 2016, as in previous years, nearly all geoscience graduates had at least one field experience or one research experience, and most participated in at least one of both types. These student experiences are essential for the development of critical research and field skills necessary for the geoscience workforce. The other critical experience in the development of an effective geoscience graduate is an internship. However, the majority of geoscience graduates do not participate in an internship experience before graduation. Internships are the best opportunities to learn the day-to-day experiences of a working geoscientist in particular industries, as well as provide an important networking opportunity. While it may be difficult to create more internship opportunities, collaborations between workforce representatives, departments, and societies are needed to look at the skills development that comes from internship opportunities in an effort to find new ways to provide internship-like experiences for current students.

While many of the trends in the report have been consistent over the past three to four years, there have been some changes in the industries that hired graduates right out of school, mostly within the oil and gas industry. Due to a downturn in the oil and gas industry, in 2016, these companies focused their hiring of new employees at the master's degree level. Concern about this downturn also negatively affected how job-seeking graduates perceived the geoscience job market. However, with oil and gas companies not hiring as many bachelor's graduates, other industries, such as the environmental services and non-profit industries, have increased rates of new hires right out of school.

AGI recognizes the importance of continuing this research study annually and is excited about the prospect of future comparisons with other countries. Soon, AGI will be able to identify those trends that are specific to U.S. graduates compared to those that are of global concern.

Contents

An Overview of the Demographics of the Participants

This year, AGI's Geoscience Student Exit Survey was made available to geoscience graduates at all traditional graduation periods (winter, spring, and, summer) during the 2015-2016 academic year, to be collectively referenced as "2016." Each spring, an email is sent to all department heads and chairs requesting their department's participation in the survey. The designated representatives from each department are periodically reminded throughout the year as to share the survey with their graduating students near the end of each semester. As incentive to participate, AGI gives the departments the data in aggregate for their graduates for their own internal assessment purposes.

AGI also enlists the help from the societies in AGI's Federation to send the survey to their student membership in the spring. In 2016, AGI received assistance with the survey distribution from the American Geophysical Union (AGU), the American Institute of Professional Geologists (AIPG), the Association for the Sciences of Limnology and Oceanography (ASLO), the Geological Society of America (GSA), and the Society of Exploration Geophysicists (SEG). These societies helped to recruit approximately 23 percent of the recent graduates that participated in the survey for 2016.

The survey was available to the winter and summer graduates for two months, and the spring graduates had three months to complete the survey. At the close of the survey, 483 graduating students from 156 geoscience schools or departments provided responses—333 bachelor's graduates, 78 master's graduates, and 70 doctoral graduates. Six states, Nevada, New Mexico, Kansas, Arkansas, New Hampshire, and Rhode Island, are not represented in this sample of geoscience graduates. The relative distribution of awarded degrees in the total sample remained the same.

The first section of the survey covered student demographics to establish an understanding of the students that graduate in the geosciences. The data remained consistent with previous years, but there was a shift in the gender distribution from 2014 and 2015. For 2016, there were more than 10 percent more male master's graduates than female master's graduates, but female doctoral graduates outnumbered male doctoral graduates by 5 percent. While the gender distribution of graduates at the three degree levels have varied over recent years, neither gender has dropped below 40 percent representation within any degree level. The survey allows for graduates to identify with other gender categories, but that option was not utilized. However, 2015 and 2016 saw increases in graduates not providing responses to some of the demographic questions, such as gender and race/ethnicity, compared to previous years. As in previous years, students indicating their citizenship as U.S. Citizen or Permanent Resident were asked to indicate their race and ethnicity. The percentage of underrepresented minorities contains African Americans, Hispanic/Latinos, Native Americans/Alaskans, and Native Hawaiians/Pacific Islanders. However, it is important to note that in this population of underrepresented minorities, the dominant group is Hispanic/Latino. The percentages of underrepresentative minority geoscience graduates continue to remain at 10 percent or less depending on the degree awarded. The age distribution of graduates in 2016 is similar to the distributions in previous years, with clear ranges of ages for graduates at each degree. Geoscience graduates over the age of 40 tend to complete master's or bachelor's degrees, likely in an attempt to gain more skills or work towards a career shift.

For the 2015 survey, recent graduates were asked to report the highest education level of their parents or guardians. Concerns have been raised that geoscience programs tend to attract students from middle and upper class families, possibly due to parental familiarity with the subject area or the high cost of the activities associated with the degree. AGI used the highest education level of parents as a proxy for inferring the socioeconomic status of geoscience graduates. In 2016, 66 percent of bachelor's graduates, 59 percent of master's graduates, and 78 percent of doctoral graduates had at least one parent with a postsecondary degree. Among the master's graduates, that is a 20 percent decrease compared to 2015, and the lowest percentage among the three degrees. While 59 percent is still high, there may be recognition among lower income students on the high value of a geoscience master's degree. This question also indicated that 10 percent of bachelor's graduates, 8 percent of master's graduates, and 7 percent of doctoral graduates were first-generation college students.

Distribution of participating graduating students and departments*

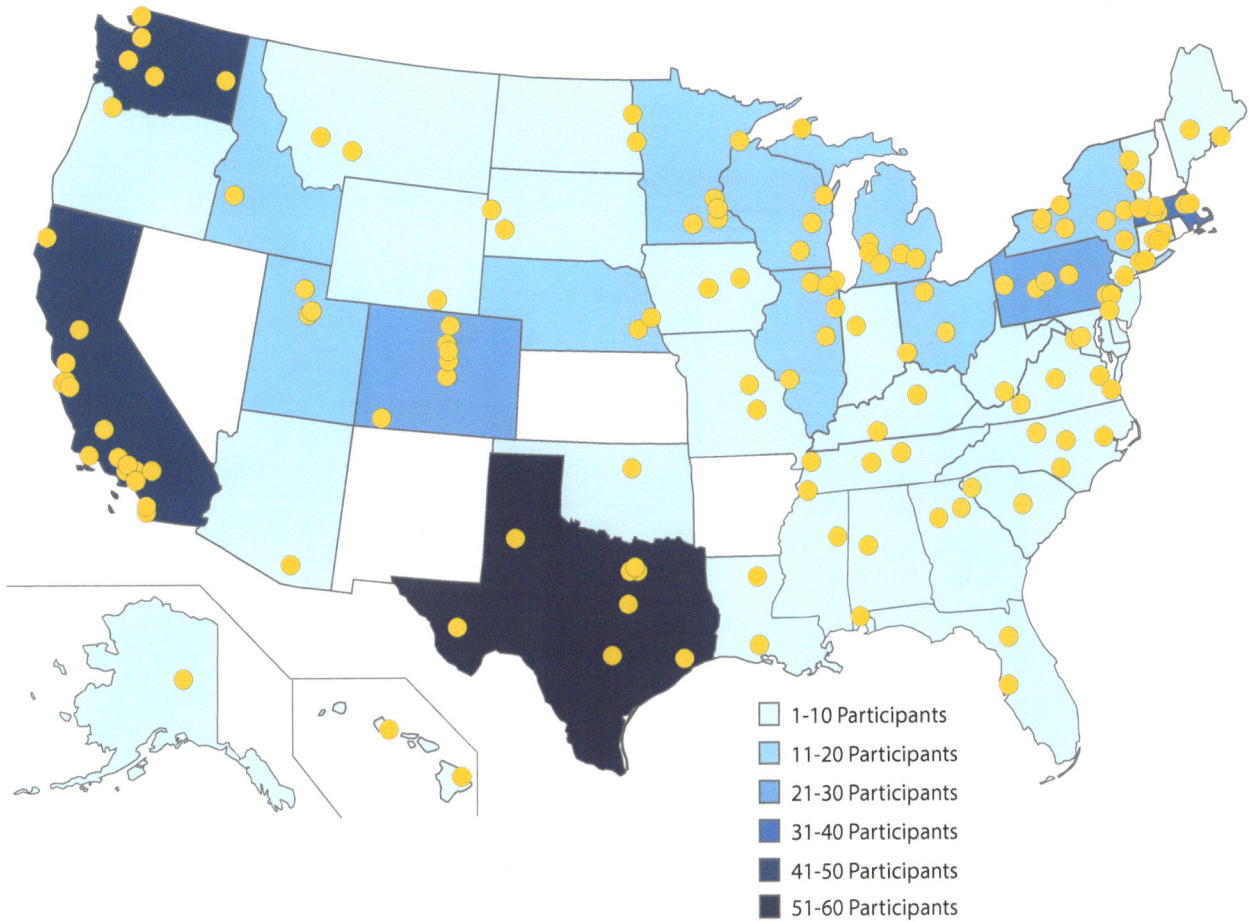

Legend:
- 1-10 Participants
- 11-20 Participants
- 21-30 Participants
- 31-40 Participants
- 41-50 Participants
- 51-60 Participants

The relative distribution by state of the universities and their graduating geoscience students across the United States that participated in the Exit Survey. *See Appendix I for list of departments

Degree received by participating graduates

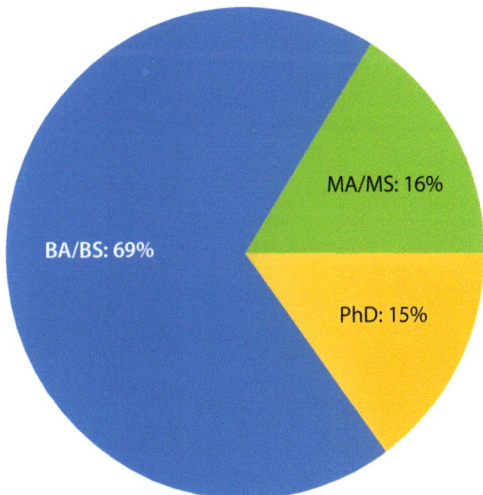

- BA/BS: 69%
- MA/MS: 16%
- PhD: 15%

Percentage of respondents within different classified institutions**

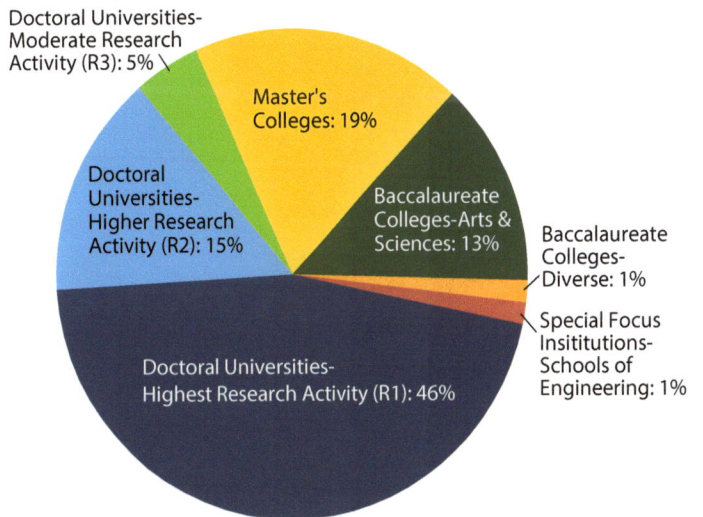

- Doctoral Universities-Moderate Research Activity (R3): 5%
- Master's Colleges: 19%
- Doctoral Universities-Higher Research Activity (R2): 15%
- Baccalaureate Colleges-Arts & Sciences: 13%
- Baccalaureate Colleges-Diverse: 1%
- Special Focus Insititutions-Schools of Engineering: 1%
- Doctoral Universities-Highest Research Activity (R1): 46%

**See Appendix II for definitions of the Carnegie University Classification System

Gender breakdown of graduates

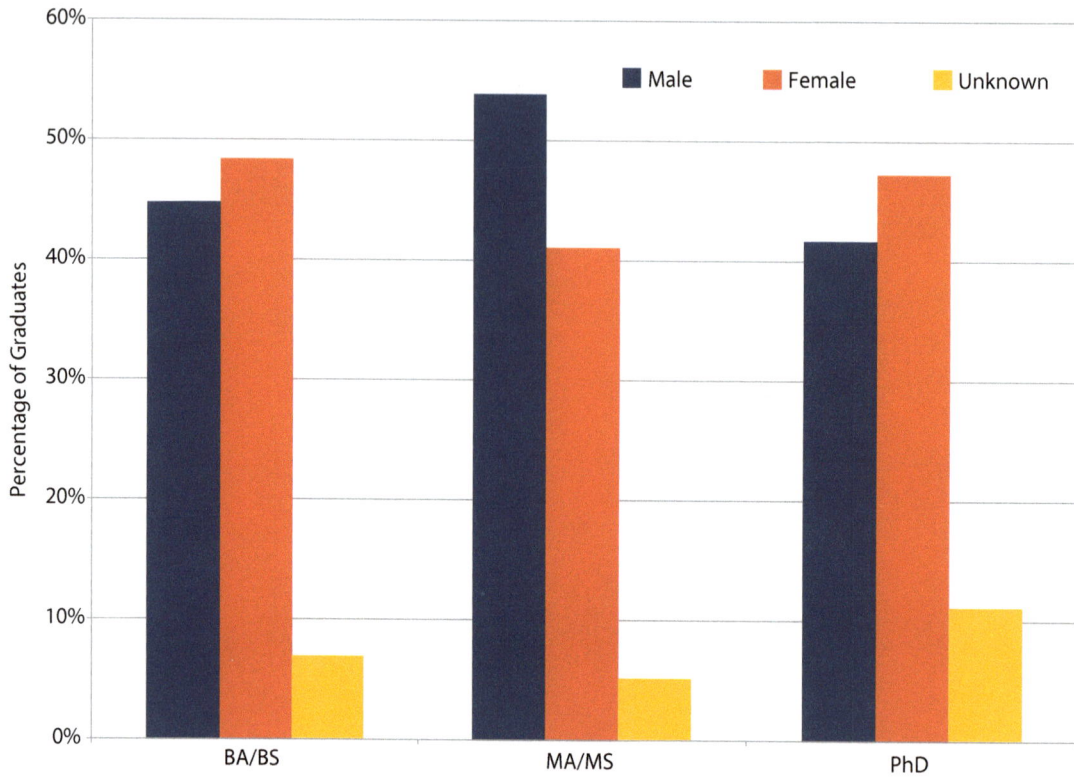

Age distribution of graduates

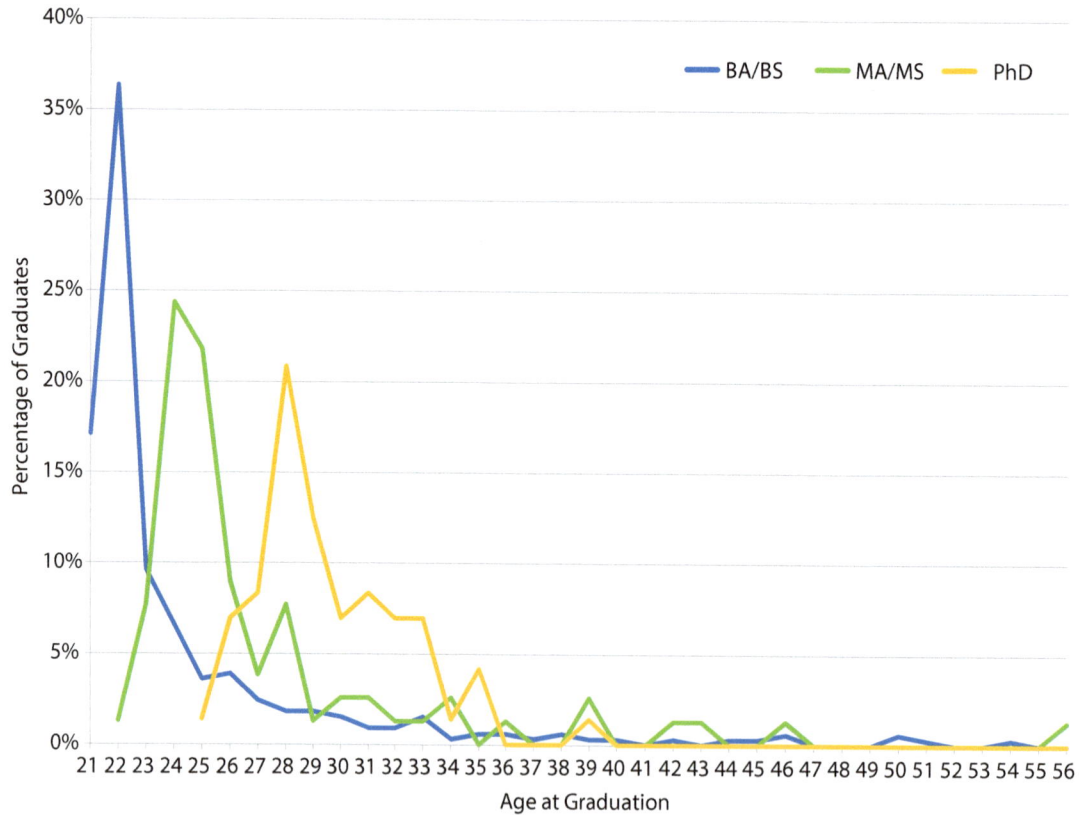

Citizenship of graduating students

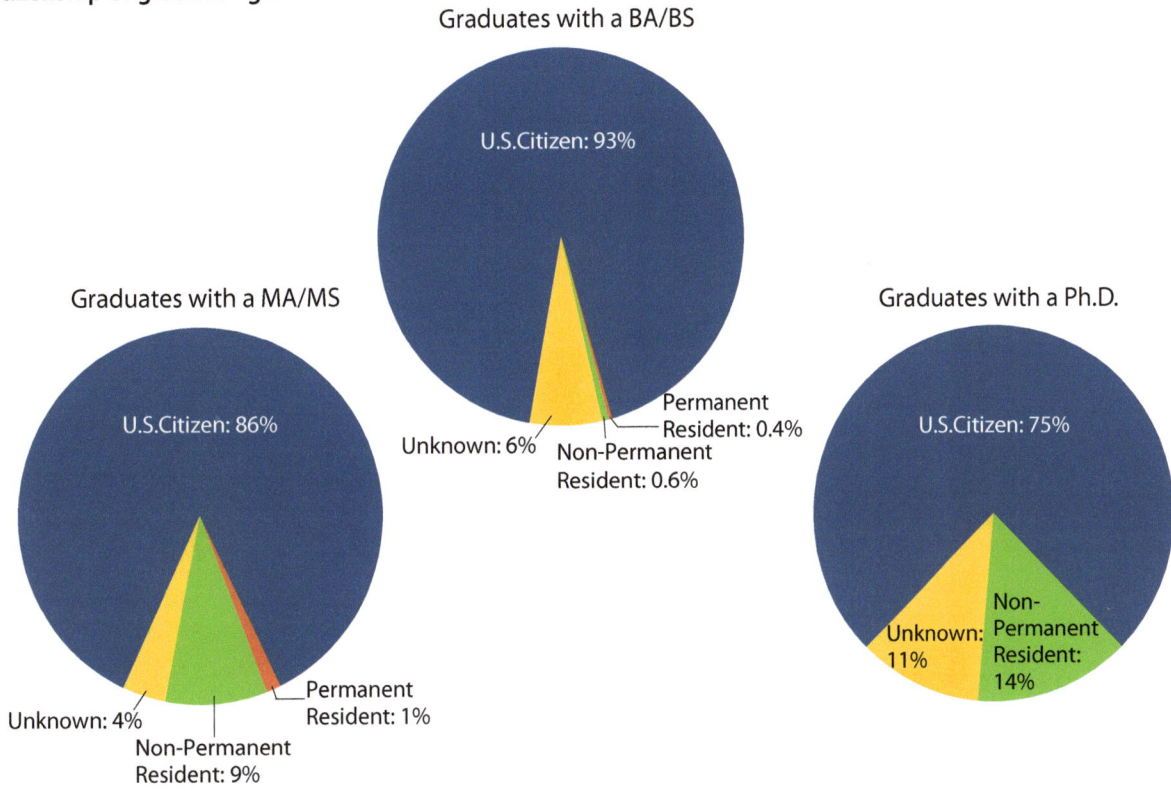

Graduates with a BA/BS

U.S.Citizen: 93%

Unknown: 6%

Permanent Resident: 0.4%

Non-Permanent Resident: 0.6%

Graduates with a MA/MS

U.S.Citizen: 86%

Unknown: 4%

Non-Permanent Resident: 9%

Permanent Resident: 1%

Graduates with a Ph.D.

U.S.Citizen: 75%

Unknown: 11%

Non-Permanent Resident: 14%

Race/ethnicity of graduating students

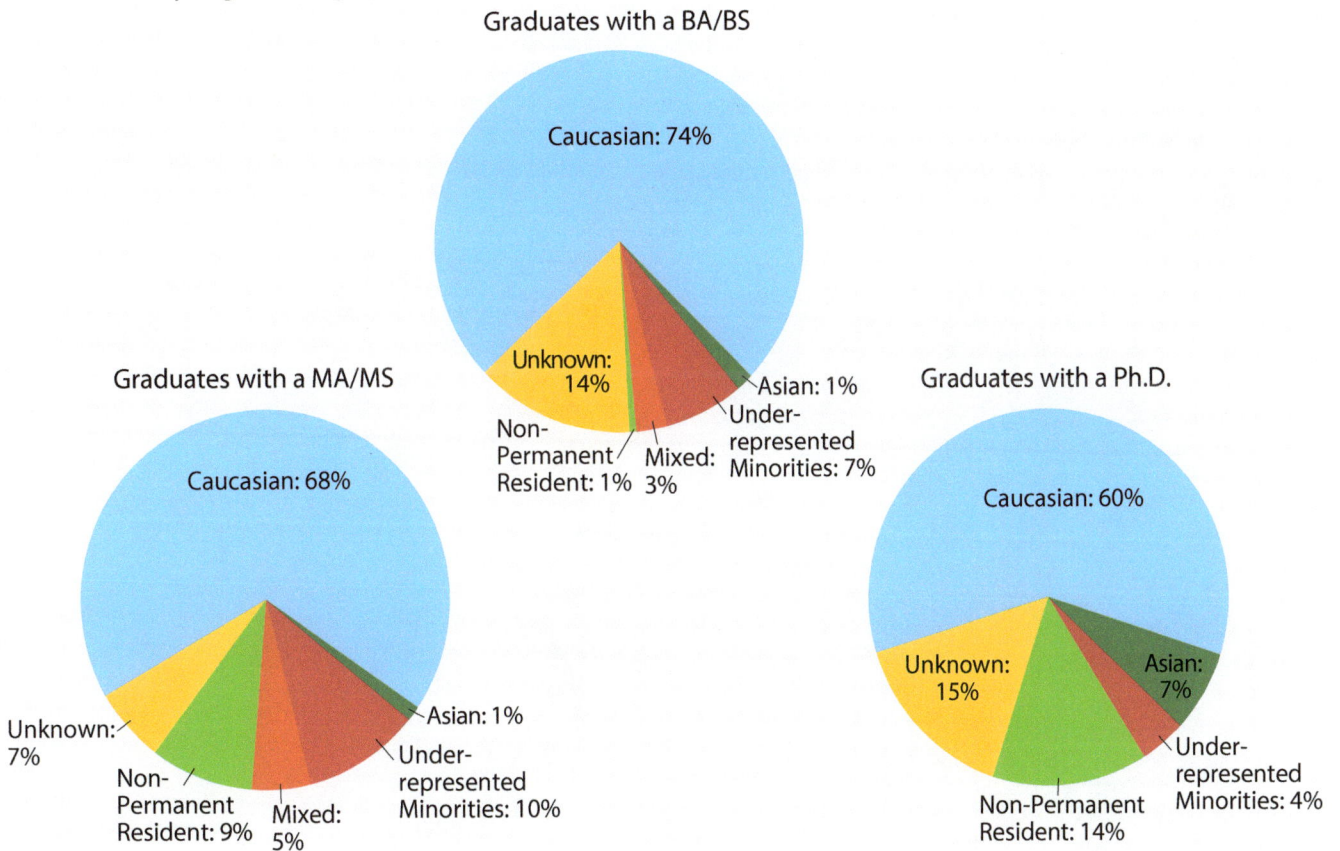

Graduates with a BA/BS

Caucasian: 74%

Unknown: 14%

Non-Permanent Resident: 1%

Mixed: 3%

Asian: 1%

Under-represented Minorities: 7%

Graduates with a MA/MS

Caucasian: 68%

Unknown: 7%

Non-Permanent Resident: 9%

Mixed: 5%

Asian: 1%

Under-represented Minorities: 10%

Graduates with a Ph.D.

Caucasian: 60%

Unknown: 15%

Non-Permanent Resident: 14%

Asian: 7%

Under-represented Minorities: 4%

Highest education level of a parent/guardian of graduates

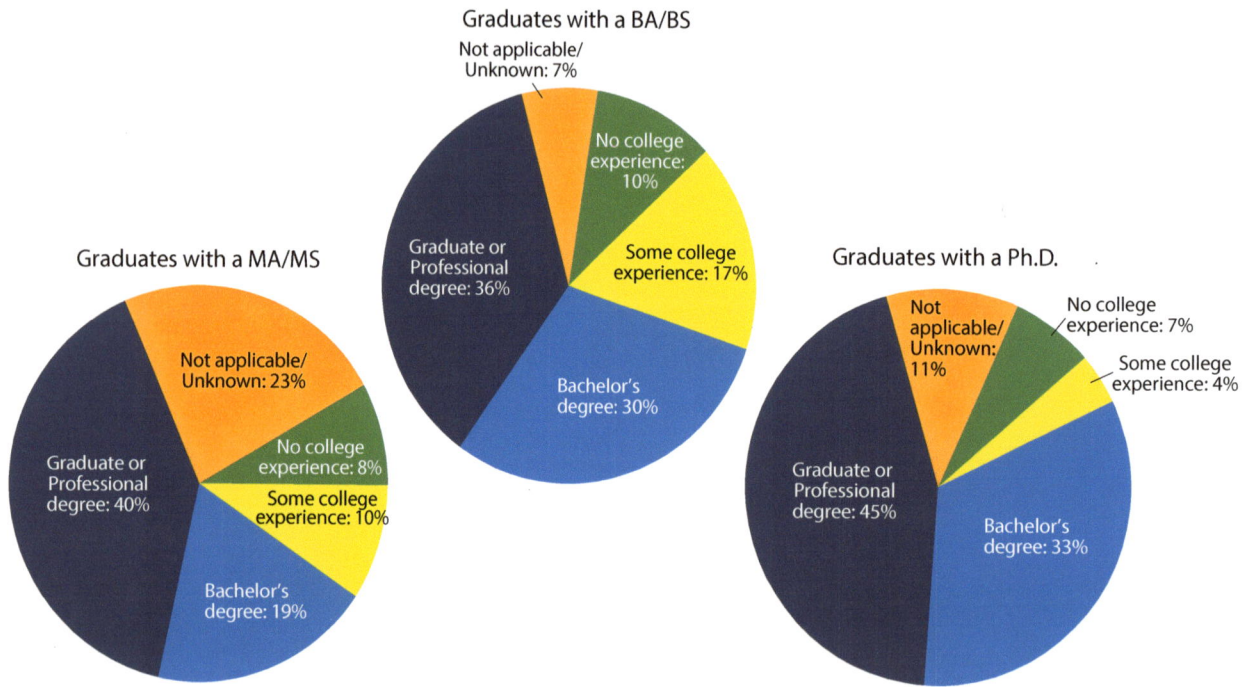

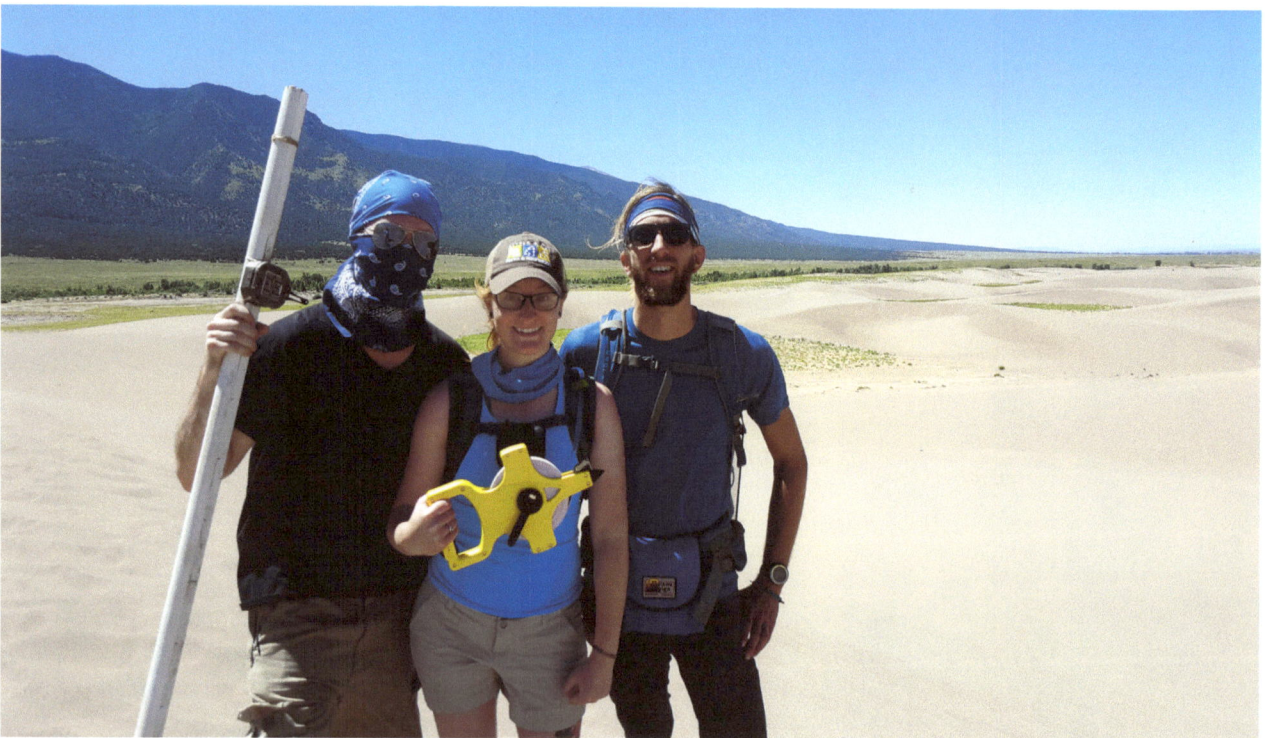

Graduates with a BA/BS

- Not applicable/Unknown: 7%
- No college experience: 10%
- Some college experience: 17%
- Bachelor's degree: 30%
- Graduate or Professional degree: 36%

Graduates with a MA/MS

- Not applicable/Unknown: 23%
- No college experience: 8%
- Some college experience: 10%
- Bachelor's degree: 19%
- Graduate or Professional degree: 40%

Graduates with a Ph.D.

- Not applicable/Unknown: 11%
- No college experience: 7%
- Some college experience: 4%
- Bachelor's degree: 33%
- Graduate or Professional degree: 45%

Image credit: Christopher DeGarmo, from AGI's 2016 Life as a Geoscientist contest

At the Great Sand Dunes National Park and Preserve Colorado, researchers mapped a dune crest, and made a dune profile.

Image credit: Isidoros Kampolis, from AGI's Life as a Geoscientist contest

Measuring the development height of a submerged tidal notch. The study was about sea-level changes from tectonic (vertical) movements in Samos Island, Greece.

Quantitative Skills and Geoscience Background of the Graduating Students

This section examines graduates' educational background, such as quantitative rigor, the role of K-12 experiences, and the importance of two-year colleges.

The graduates were asked to select all of the quantitative courses they have taken at a two-year or four-year institution. Consistently over the past three years, the majority of geoscience graduates, regardless of degree, complete Calculus II as their highest quantitative course. While a decrease in the level of upper level math courses is visible among all degree levels, the higher the degree completed, the more likely the graduate has taken at least one quantitative course beyond Calculus II. Some of these higher quantitative courses are likely taken by graduate students, but this does raise the question if an increasing proportion of graduates that take courses beyond Calculus II are entering graduate programs after taking these high level quantitative courses as an undergraduate. The number of geoscience bachelor's graduates far exceeds graduation rates for master's and doctoral degrees, so it is likely that the small number of bachelor's graduates that have taken the higher quantitative courses are the same students that move on to earn a graduate degree. Due to the complex physical nature of studying earth systems and geosciences, it is essential for recent graduates to have experience in coursework, like Linear Algebra and Differential Equations, in order to effectively understand areas such as fluid dynamics and multivariate systems. When looking at the participation in quantitative courses by the type of institution, most of the participation in these higher quantitative courses occurred at doctorate granting institutions relative to rates at liberal arts colleges. If a large portion of graduate students with experience in courses, such as Differential Equations and Linear Algebra, gain this experience during their undergraduate degree, are liberal arts graduates starting at a disadvantage when applying for graduate school? Participation rates in statistics courses has also raised some concern. Over the past three years, approximately 40 percent of geoscience doctoral graduates did not take a statistics course during their postsecondary education. Participation in statistics by doctoral graduates should be much closer to 100 percent. It is essential for the geoscience academic workforce to have a solid background in statistics in order to successfully interpret and evaluate the veracity of published research and add to the research within their fields.

Students were asked if they took an earth or environmental science course in high school and if they attended a two-year college for at least one semester before receiving their degree. From 2013-2015, approximately half the graduates took an earth or environmental course in high school. In 2016, however, those percentages dropped below 50 percent for all degree levels. While these courses may or may not be the reason a student majors in the geosciences, high school exposure to earth science can create an interest in the subject area, as well as a comfort level in the subject entering into an introductory geoscience course in college. Two-year college continues to be an important recruitment venue for the geosciences, with nearly one third of bachelor's graduates and one quarter of master's graduates attending a two-year college for at least a semester. Two-year colleges are becoming a viable and necessary option for many students to begin their postsecondary education. More collaborations and agreements between two-year and four-year institutions are needed to help students with the transition and completion of the bachelor's degree.

Image credit: Zsuzsanna Thoth, from AGI's 2016 Life as a Geoscientist contest

Sketching important structural features, Gogama, ON, Canada

Quantitative skills and knowledge gained while working towards degree

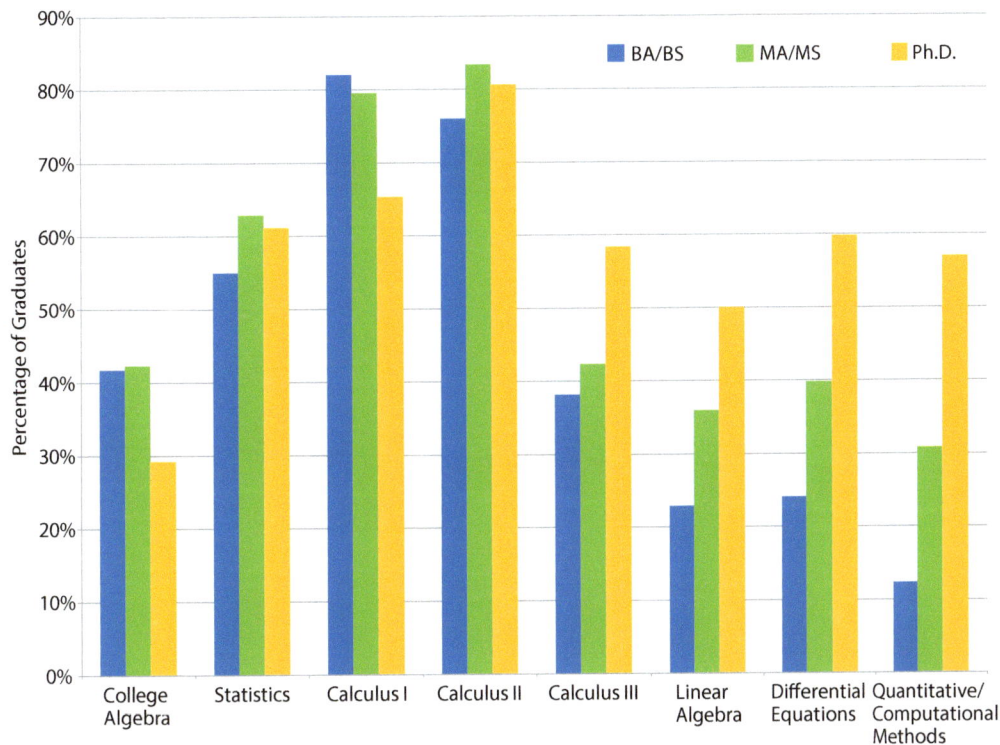

Quantitative skills and knowledge gained based on university classification**

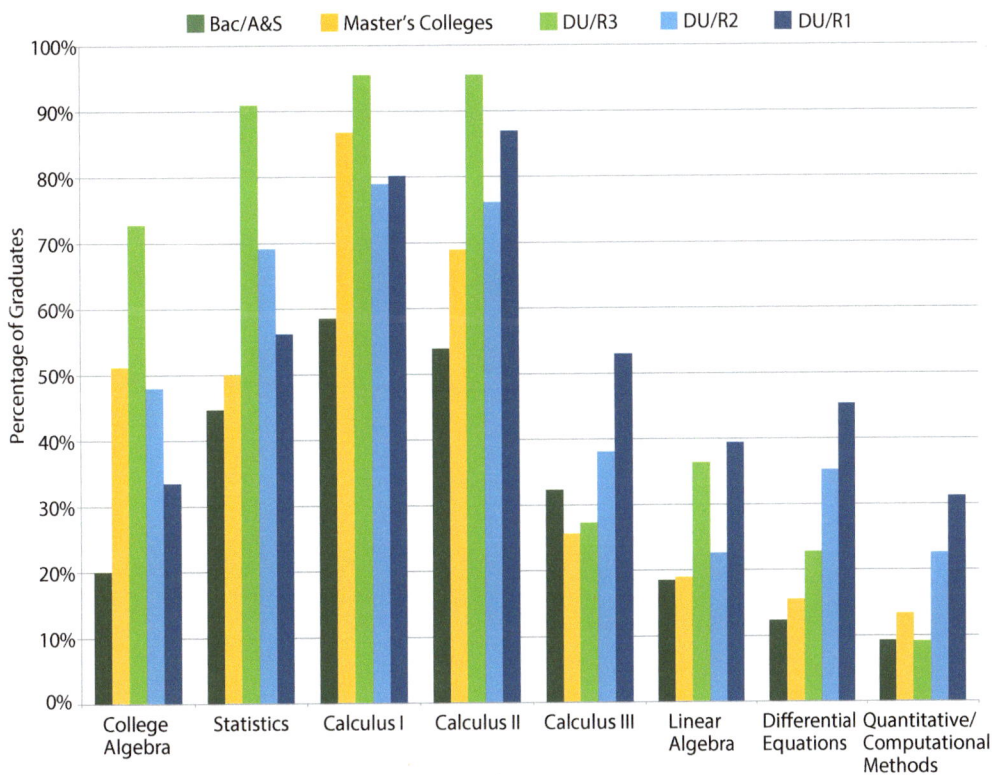

**See Appendix II for defintions of the Carnegie University Classification System

Quantitative skills and knowledge gained while working towards degree by gender

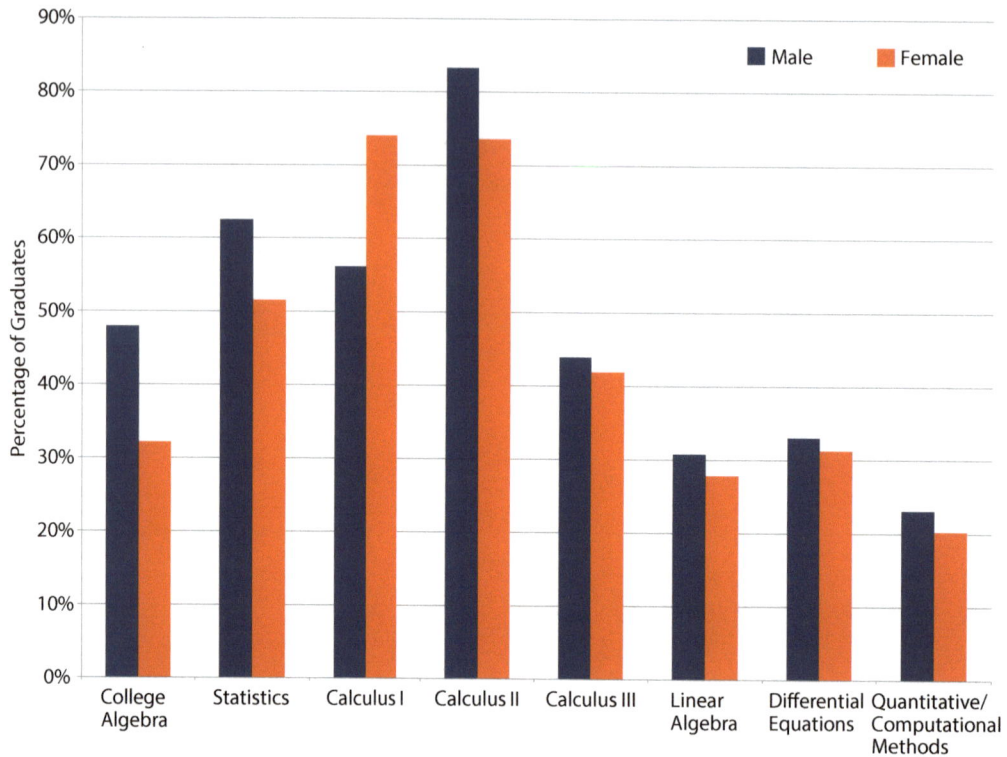

Percentage of graduates taking supplemental science courses

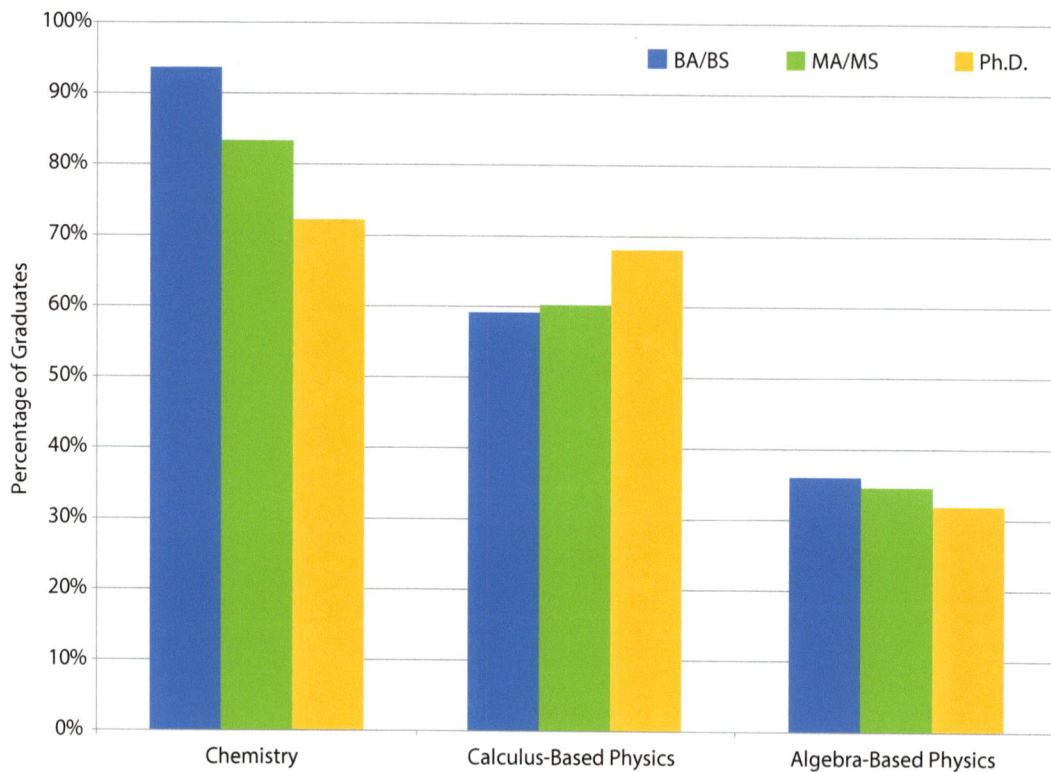

Graduates who took an earth science course in high school

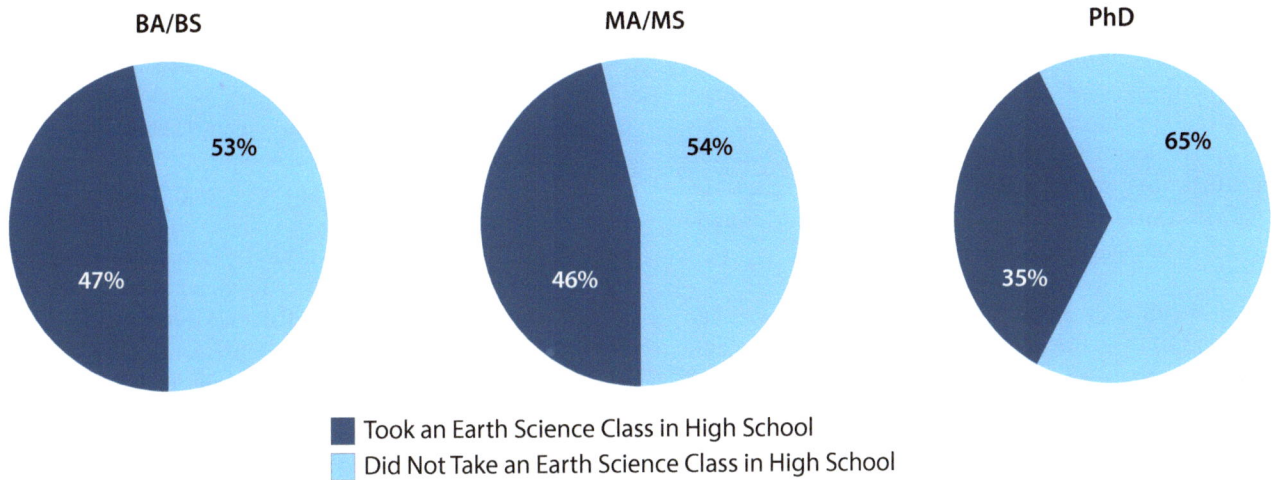

BA/BS

53%

47%

MA/MS

54%

46%

PhD

65%

35%

■ Took an Earth Science Class in High School
□ Did Not Take an Earth Science Class in High School

Graduates who attended a two-year college for at least 1 semester and took a geoscience course

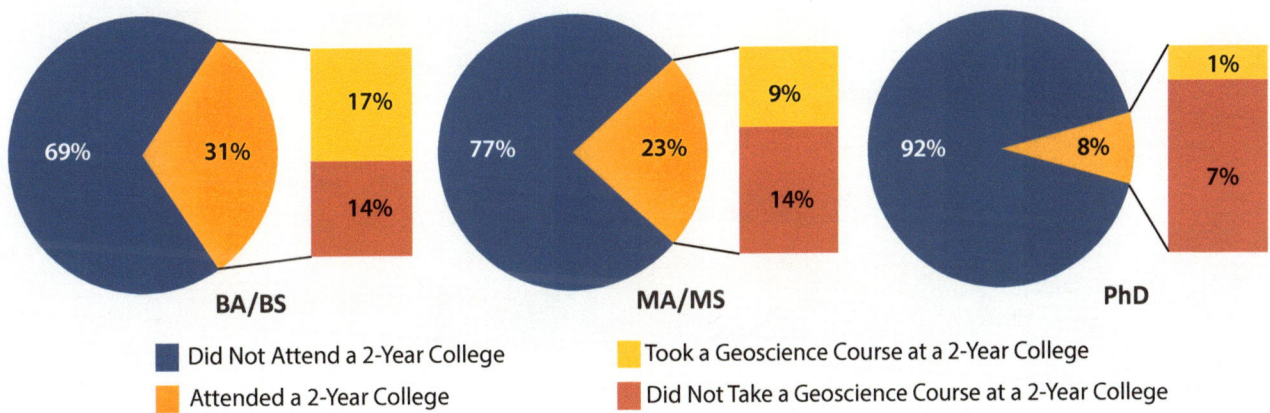

69% 31%

17%

14%

BA/BS

77% 23%

9%

14%

MA/MS

92% 8%

1%

7%

PhD

■ Did Not Attend a 2-Year College
■ Attended a 2-Year College

■ Took a Geoscience Course at a 2-Year College
■ Did Not Take a Geoscience Course at a 2-Year College

Choosing Geoscience as a Major

raduates were asked which geoscience field they were pursuing with their degree, as well as the fields associated with any other postsecondary degrees. The chosen degree fields demonstrate the variety of disciplines related to the geosciences. Geology continues to be the most popular degree among undergraduates with students tending to specialize in different fields upon entering graduate school.

Consistently, the majority of graduates at the bachelor's and master's levels chose to major in the geosciences at some point during their undergraduate education. In 2014 and 2015, most doctoral graduates chose to major in the geosciences either after completing an undergraduate degree or before beginning college; but in 2016, most doctoral graduates chose the geosciences either during their undergraduate degree or after completing an undergraduate degree. The timing of the choices of bachelor's and masters' students demonstrate the importance of the introductory geoscience courses for recruitment into the majors, but the doctoral graduates highlight the interdisciplinary nature

of the geosciences due to the high percentage of doctoral graduates that change their majors to the geoscience after receiving and undergraduate degree. Bachelor's or master's graduates in other sciences, particularly chemistry or physics, can easily transfer to the geosciences for future degrees with their strong physical science background.

The graduates were asked to briefly explain their reasoning for majoring in the geosciences. As in previous years, the majority of graduates at all degree levels indicated the intellectual engagement of the geosciences as the reason for choosing their major. The comments included reasons related to the interdisciplinary nature of the geosciences, passion for certain degree fields, and the inherent interest of being outdoors and asking questions of the environment around them. In 2016, nearly all the doctoral graduates' responses to this question were related to their interest in the subject. Bachelor's and master's graduates also included reasons related to connections with faculty and staff in the departments, familial influence, and career options in the geosciences.

Time when students decide to major in the geosciences

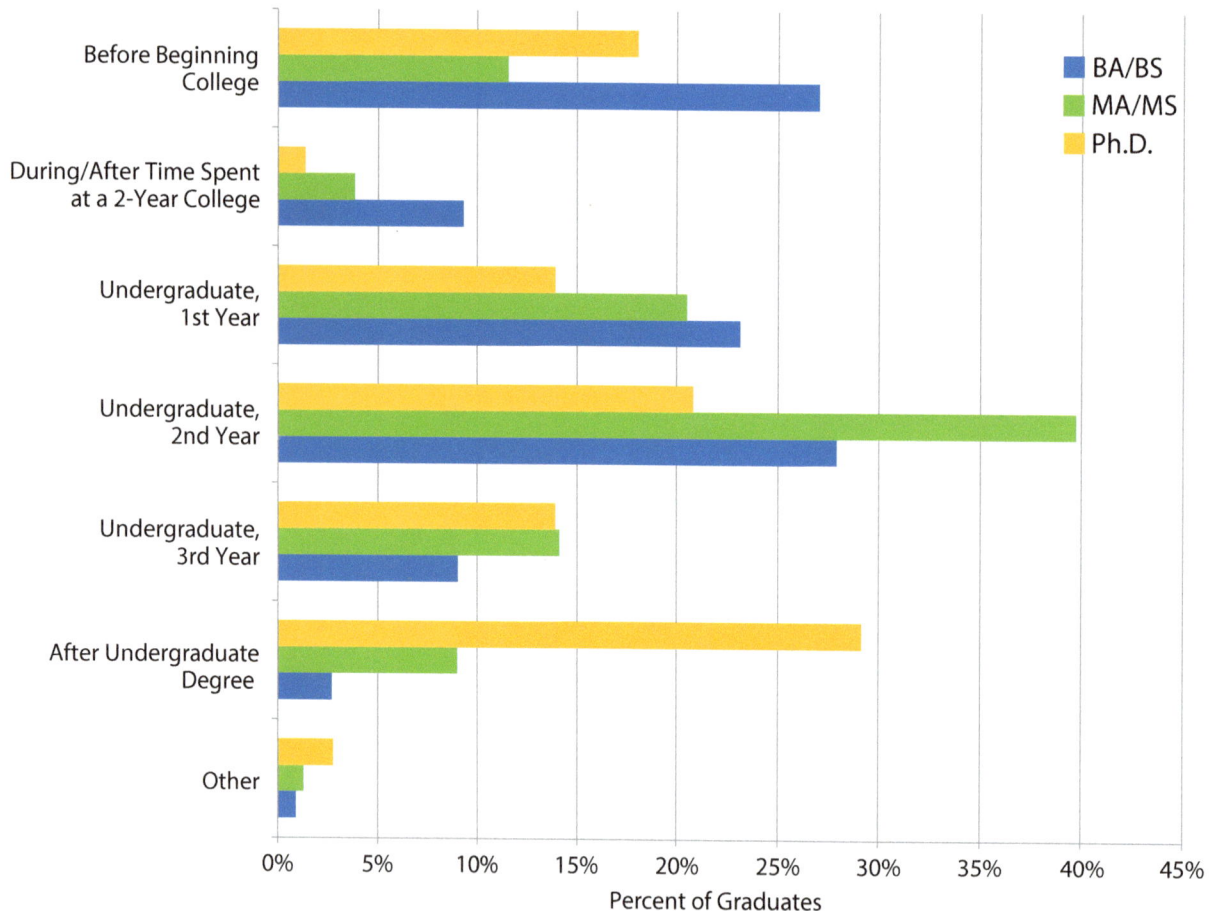

Percent of Graduates

Chosen geoscience degree fields

Bachelor's Degree Graduates

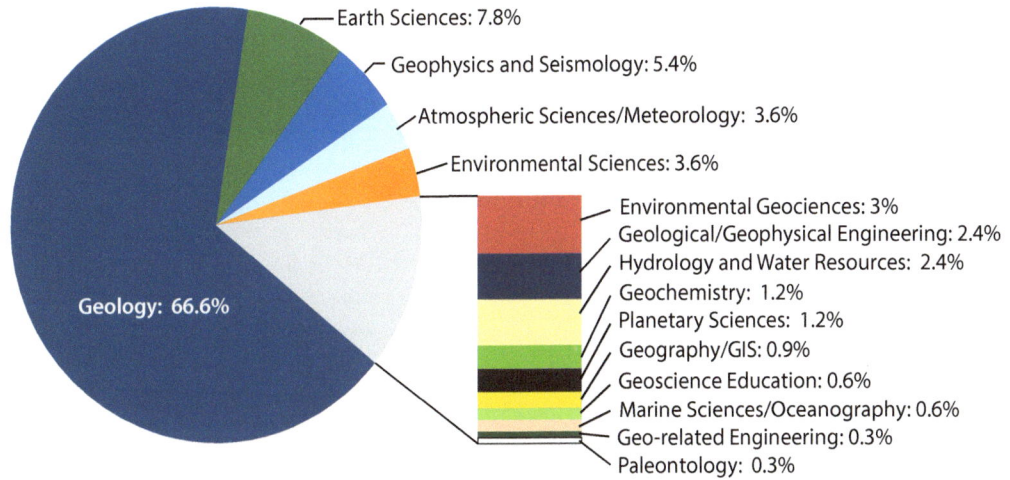

- Earth Sciences: 7.8%
- Geophysics and Seismology: 5.4%
- Atmospheric Sciences/Meteorology: 3.6%
- Environmental Sciences: 3.6%
- Geology: 66.6%
- Environmental Geocences: 3%
- Geological/Geophysical Engineering: 2.4%
- Hydrology and Water Resources: 2.4%
- Geochemistry: 1.2%
- Planetary Sciences: 1.2%
- Geography/GIS: 0.9%
- Geoscience Education: 0.6%
- Marine Sciences/Oceanography: 0.6%
- Geo-related Engineering: 0.3%
- Paleontology: 0.3%

Master's Degree Graduates

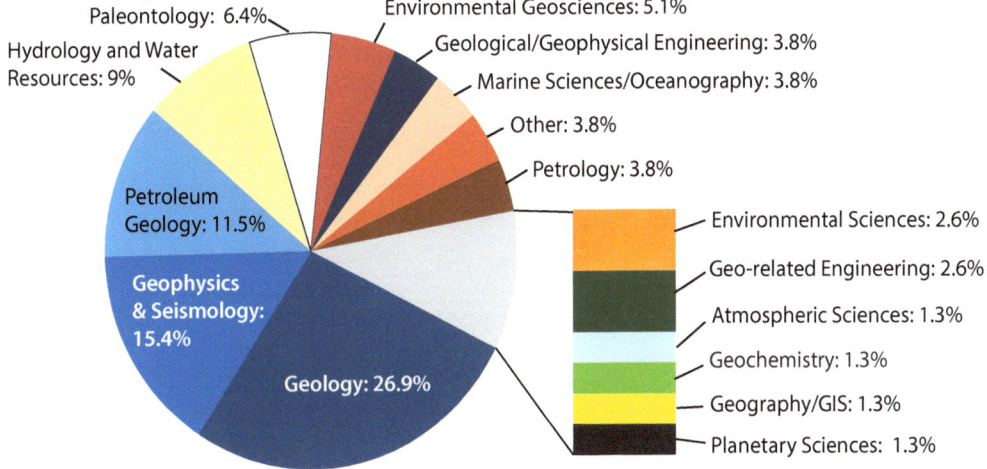

- Paleontology: 6.4%
- Hydrology and Water Resources: 9%
- Petroleum Geology: 11.5%
- Geophysics & Seismology: 15.4%
- Geology: 26.9%
- Environmental Geosciences: 5.1%
- Geological/Geophysical Engineering: 3.8%
- Marine Sciences/Oceanography: 3.8%
- Other: 3.8%
- Petrology: 3.8%
- Environmental Sciences: 2.6%
- Geo-related Engineering: 2.6%
- Atmospheric Sciences: 1.3%
- Geochemistry: 1.3%
- Geography/GIS: 1.3%
- Planetary Sciences: 1.3%

Doctoral Degree Graduates

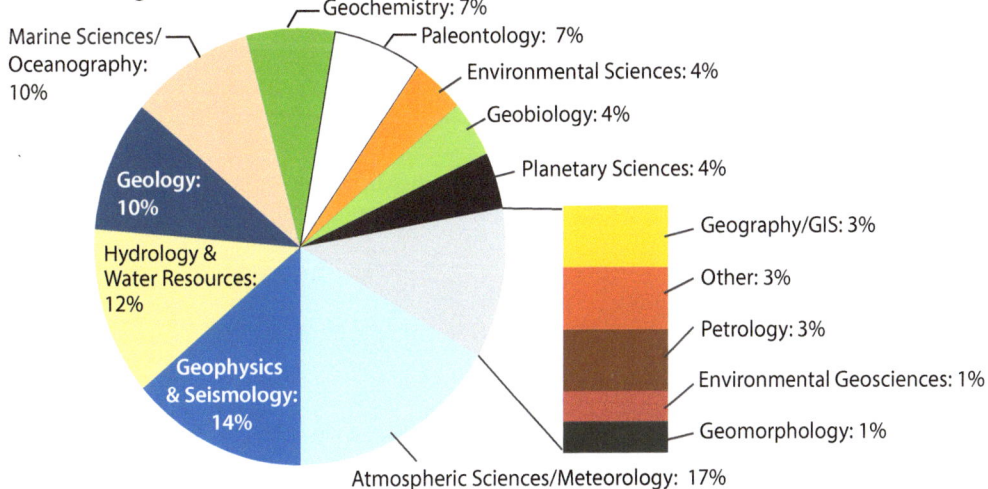

- Geochemistry: 7%
- Paleontology: 7%
- Environmental Sciences: 4%
- Geobiology: 4%
- Planetary Sciences: 4%
- Marine Sciences/Oceanography: 10%
- Geology: 10%
- Hydrology & Water Resources: 12%
- Geophysics & Seismology: 14%
- Atmospheric Sciences/Meteorology: 17%
- Geography/GIS: 3%
- Other: 3%
- Petrology: 3%
- Environmental Geosciences: 1%
- Geomorphology: 1%

Ancillary Factors Supporting the Degree

Graduates were asked about their experiences while working towards their degree. In 2016, there were low participation rates in internships by graduates, particularly at the bachelor's and doctorate levels. This has been a consistent trend since the start of AGI's Exit Survey in 2013. Master's graduates tend to have higher internship participation rates, most likely due to their alignment towards preparation for their professional career. However in 2015 and 2016, 40 percent of master's graduates did not participate in an internship before graduation. To investigate these low participation rates in internships further, the graduates are asked for the number of internship applications they submitted and the resources used to find internship announcements. While some graduates attempted to get an internship, there are still over 40 percent of bachelor's and doctoral graduates that did not submit an application.

The data suggest there might be two different issues related to the low participation rates in internships: the availability of internships to students and trouble finding such opportunities. Concern has been raised that there are not enough internship like opportunities for geoscience students at all degree levels, and this is supported by the percentage of graduates that applied for internships but were unable to secure one. Industry representatives have discussed the difficulties in providing these opportunities, such as the cost of an intern and the time spent training a temporary employee. The low participation rates may also be due to difficulty in finding appropriate opportunities than can fit into an already packed degree program. There is not a centralized listing of internships for geoscience students, and departments may not be the best source for finding these opportunities. According to the graduates, most internship announcements were found through internet searches. It is essential that the geosciences community recognize the importance of internship activities for students' professional development. Consideration is needed for ways to promote and provide this type of professional development to current students at all degree levels.

Graduates were asked about their usage of financial aid while working on their degree. In 2016, 81 percent of bachelor's graduates, 88 percent of master's graduates, and 94 percent of doctoral graduates used at least one form of financial aid to complete their degree. Consistent with past years, in 2016, bachelor's graduates depended on student loans and federal grants to help pay for school, and master's and doctoral graduates tended to depend on research and teaching assistantships to help pay for school. Doctoral graduates also benefitted from department grants and external grants more than the other degree levels. It is important to note that 36 percent of master's graduates and 17 percent of doctoral graduates used student loans to help pay for their degree. It is a bit of a misunderstanding that all graduate students have their degree paid for in full. Many schools may pay the tuition bill, but the attached fees are often up to the student to pay on their own. Also, due to increased pressure by universities to increase enrollments, some graduate programs can't afford to help financially support all of their students.

Graduates were also asked about their involvement with geoscience membership organizations. AGI is a federation of 51 geoscience societies, including the American Geophysical Union, the American Institute of Professional Geologists, the Association for the Sciences of Limnology and Oceanography, the Geological Society of America, and the Society of Exploration Geophysicists. Professional societies can be useful tools for students and recent graduates to be successful as early-career geoscientists. While participation rates in geoscience organizations at all degree levels were relatively high, it appears that most students are not affiliated with one of AGI's geoscience societies, which is surprising, especially at the doctorate level. However, participants in the survey can respond yes to their participation in geoscience organizations but choose not to list the organizations with which they are affiliated. Participation rates in AGI societies are likely higher than reported. Most students that do participate in a geoscience organization claim to be affiliated with a department level organization. This shows the importance of these department organizations to provide academic support and social connections for students in the program.

Number of internships held by graduating students

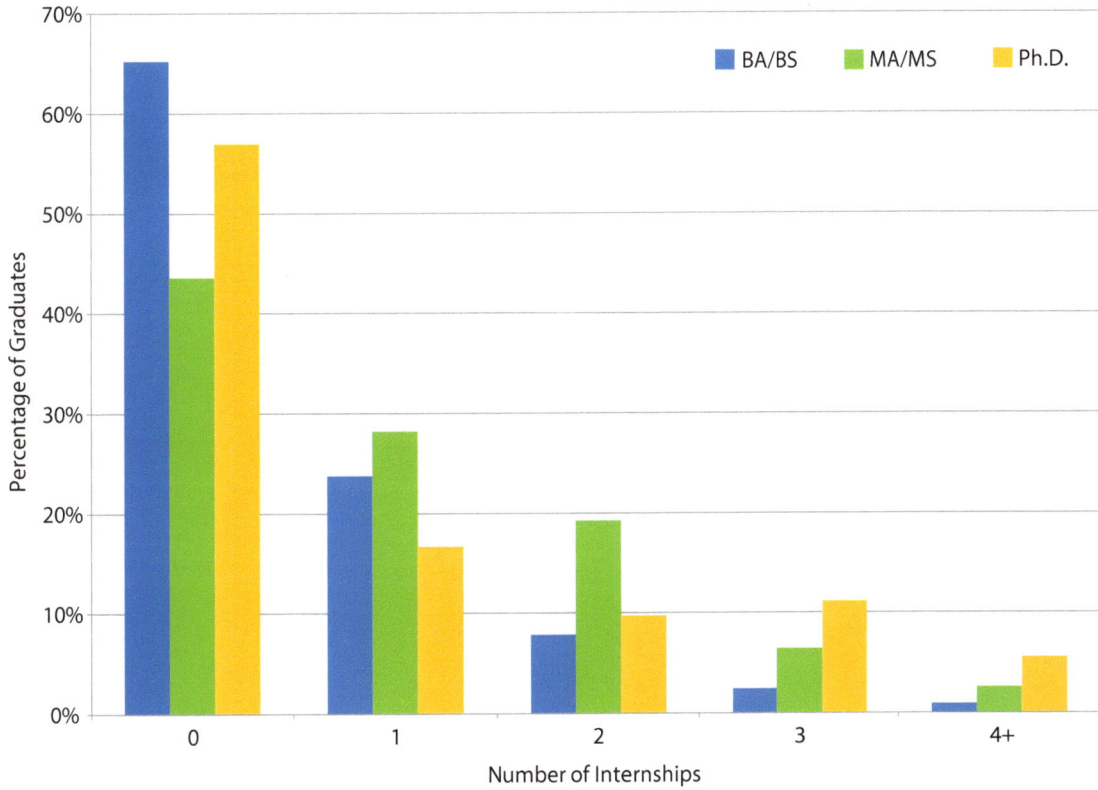

Internship applications completed by graduates

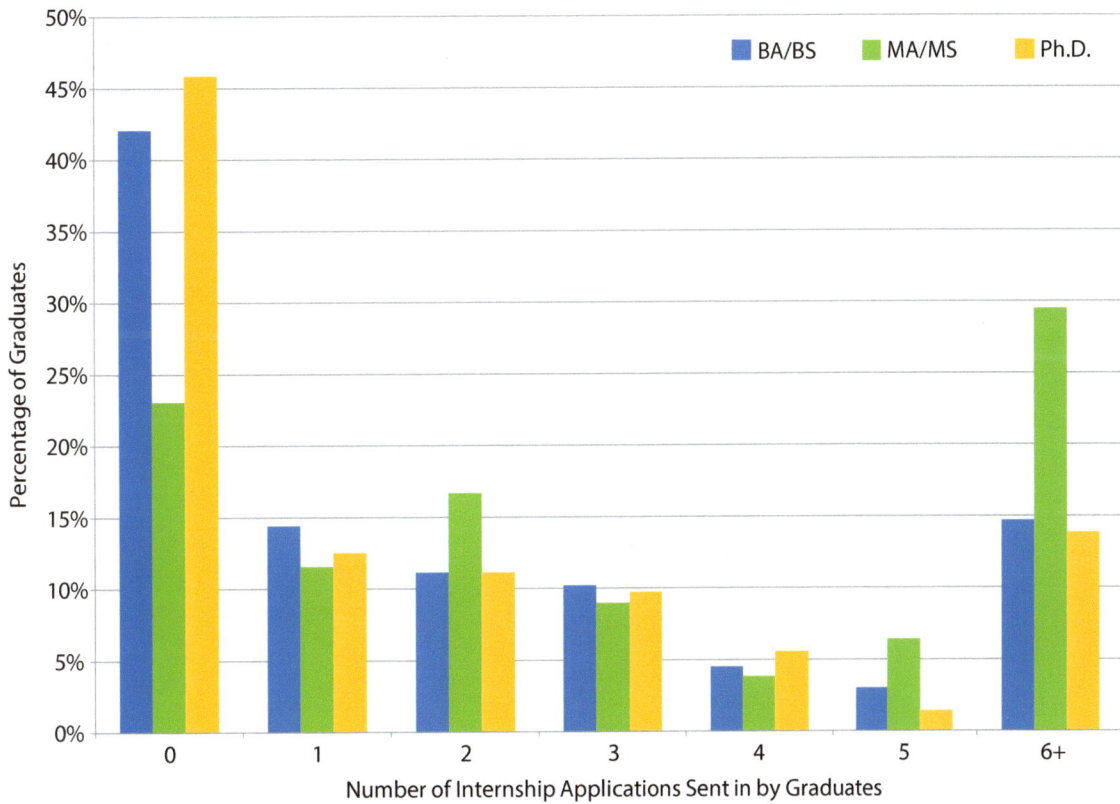

Resources used to find internship announcements

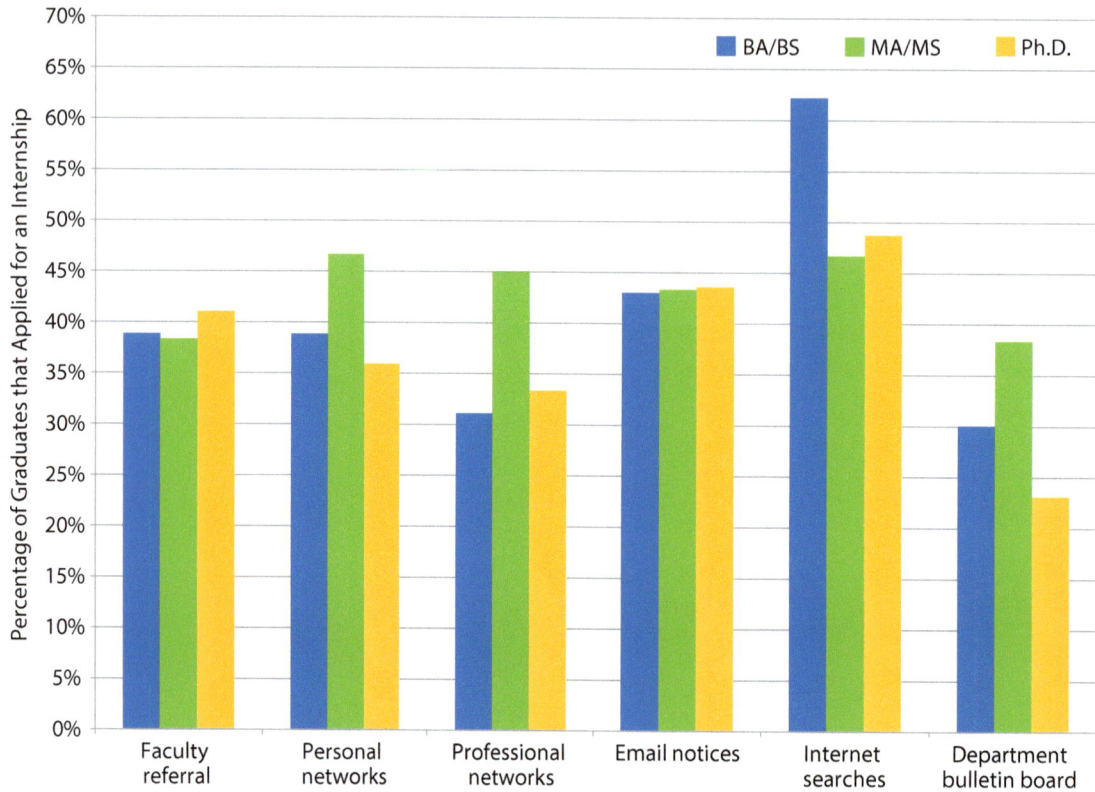

Types of financial aid used by graduating students while working towards a degree

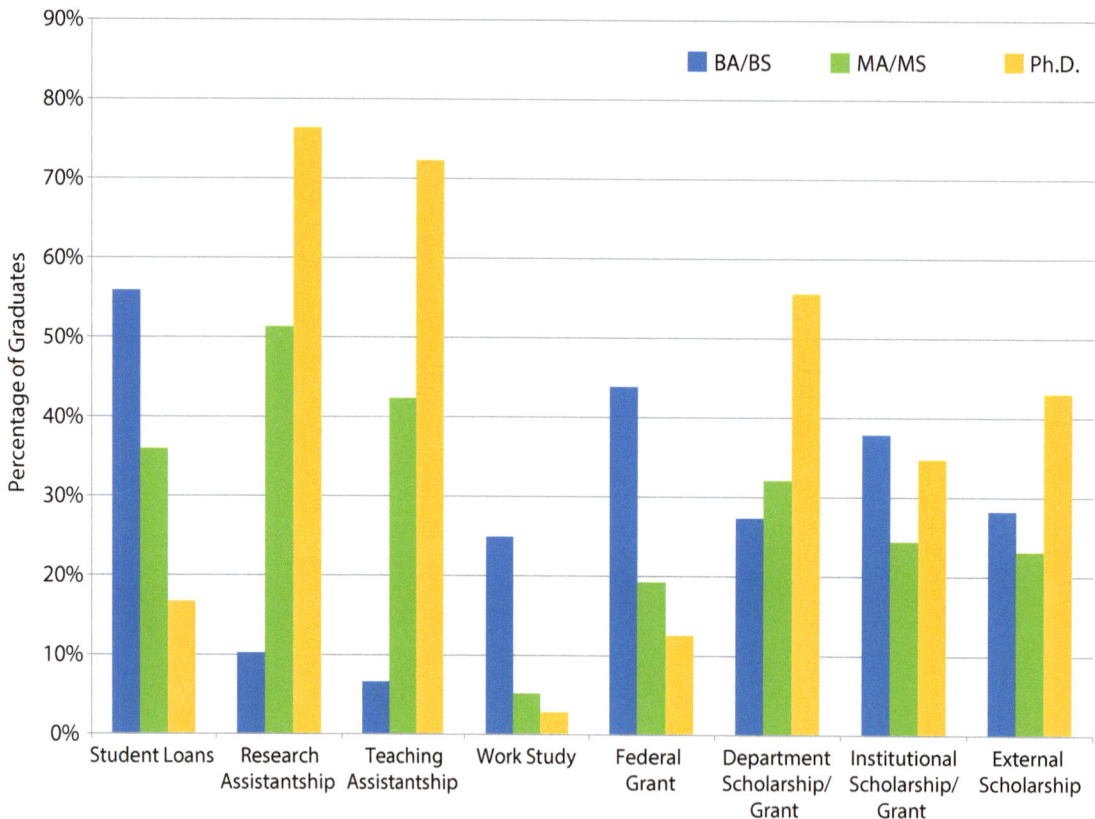

Participation in geoscience organizations

	BA/BS	MA/MS	Ph.D.
Associated with a geoscience-related club/organization	71%	82%	60%
Participated in department-level geoscience club	52%	49%	29%
Member of an AGI Member Society	22%	47%	17%
Member of an Honor Society	7%	8%	7%

Average GPA

	BA/BS	MA/MS	Ph.D.
Average years to degree completion	3.54	1.99	4.81
Average overall GPA	3.33	3.78	3.84
Average geoscience GPA	3.44	3.76	3.85

Image credit: Justin Lawrence, from AGI's 2016 Life as a Geoscientist contest

McMurdo Sound, Antarctica. Installing the first critical pieces of our field camp (bathrooms!) for a season of exploring beneath the McMurdo Ice Shelf.

Field Experiences

Clear definitions were provided to distinguish between field camp, field courses, and field experiences. A field camp was defined as an academic program lasting four or more weeks that is primarily focused on field tools and methods. Because field camp is typically an experience only taken once, this question covers the graduates' entire postsecondary education. A field course was defined as a course with a field component primarily covering field methods and experimentation that utilized at least half of the total class time. A field experience was defined as any course that contained a field component, such as a field trip, field work, or other time in the field, that is not included in the definitions for field camp or field courses.

In 2016, about 5 percent of geoscience graduates did not participate in any field experiences while working on their degree. Since 2014, there has been a 16 percent drop in the participation rate of doctoral graduates completing a field camp experience with 44 percent participation in 2016. Field camp participation rates among bachelor's and master's graduates have stayed relatively the same from 2014-2016. Participation in field courses decreased among bachelor's graduates by 8 percent in 2016 compared to bachelor's graduates in 2015, and participation in field experiences among master's and doctoral graduates decreased by 17 percent and 16 percent respectively in 2016 compared to 2015. It's not clear why there are these decreases in participation in field courses and field experiences among the graduates because most geoscience programs provide these opportunities throughout the curriculum. In fact, participation in field experiences was high regardless of the type of institution. However, availability of field camp opportunities continue to be low among liberal arts colleges and master's colleges.

Field experiences, particularly those that teach effective field skills, for geoscience students are essential for their education and training for the workforce. Since 1996, participation rates in field camp have soared to a point where many annual field camps are at capacity each year. Because of this overall increased participation in field camp across the United States, many employers expect recent graduates to have the necessary field skills developed through a field camp course. For those recent graduates that do not have access to a field camp, many of those necessary skills can be gained through field courses, as long as participation in these courses is encouraged by the geoscience programs.

Student participation in field experiences based on university classification

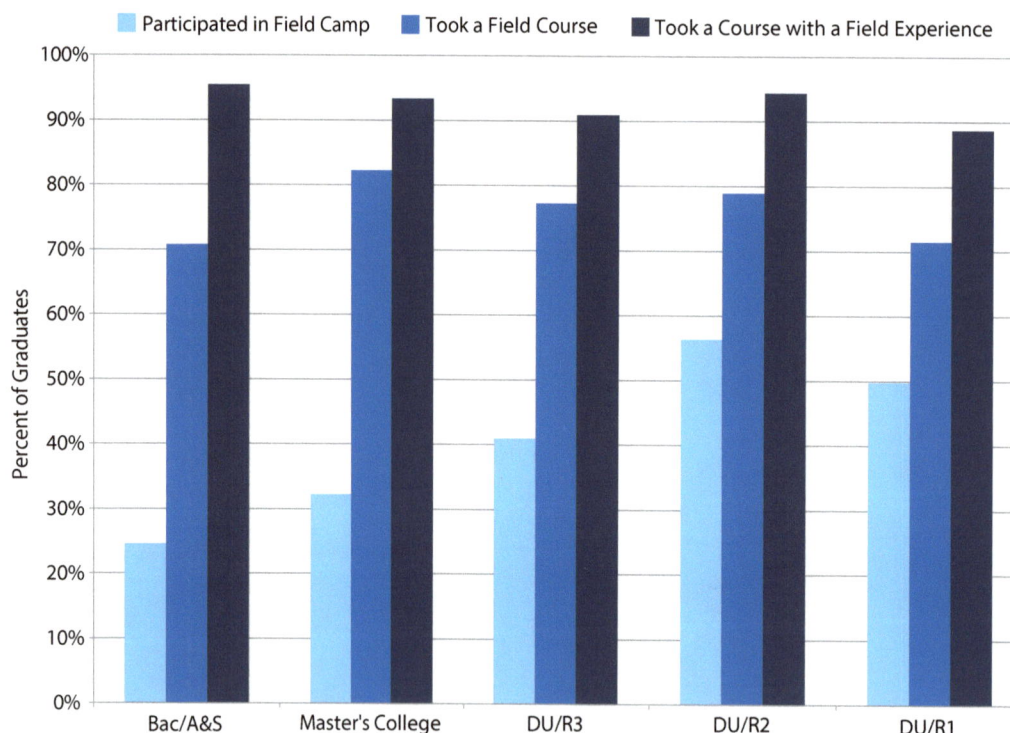

**See Appendix II for definitions of the Carnegie University Classification System

Graduating students who have participated in field camp

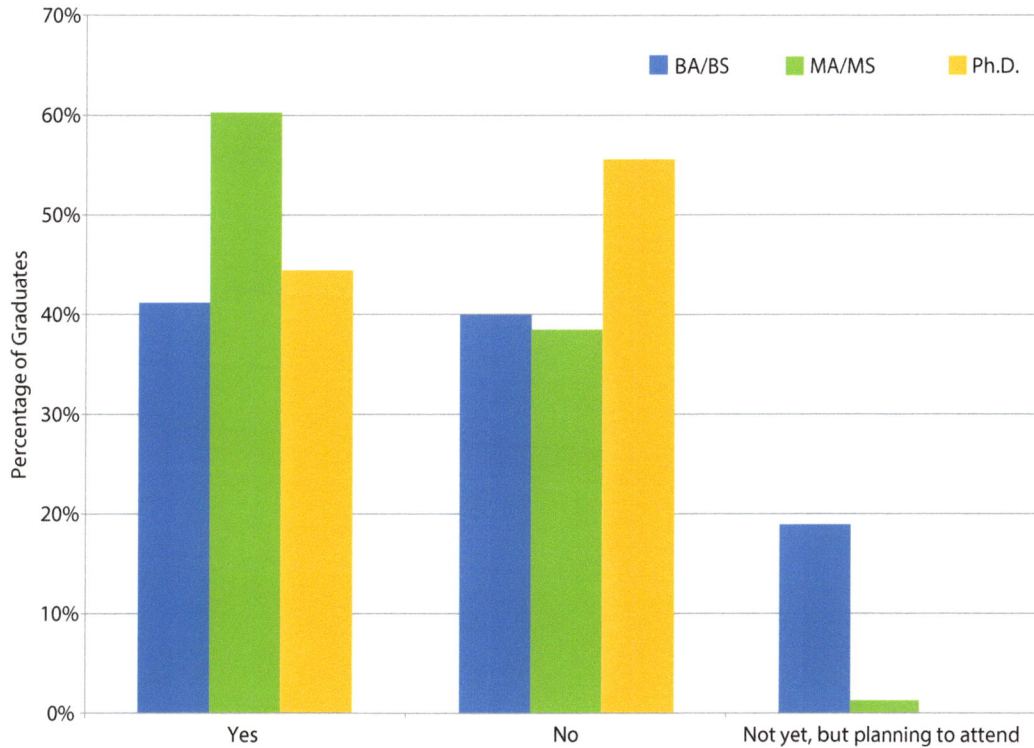

Graduating students who have participated in field camp by gender

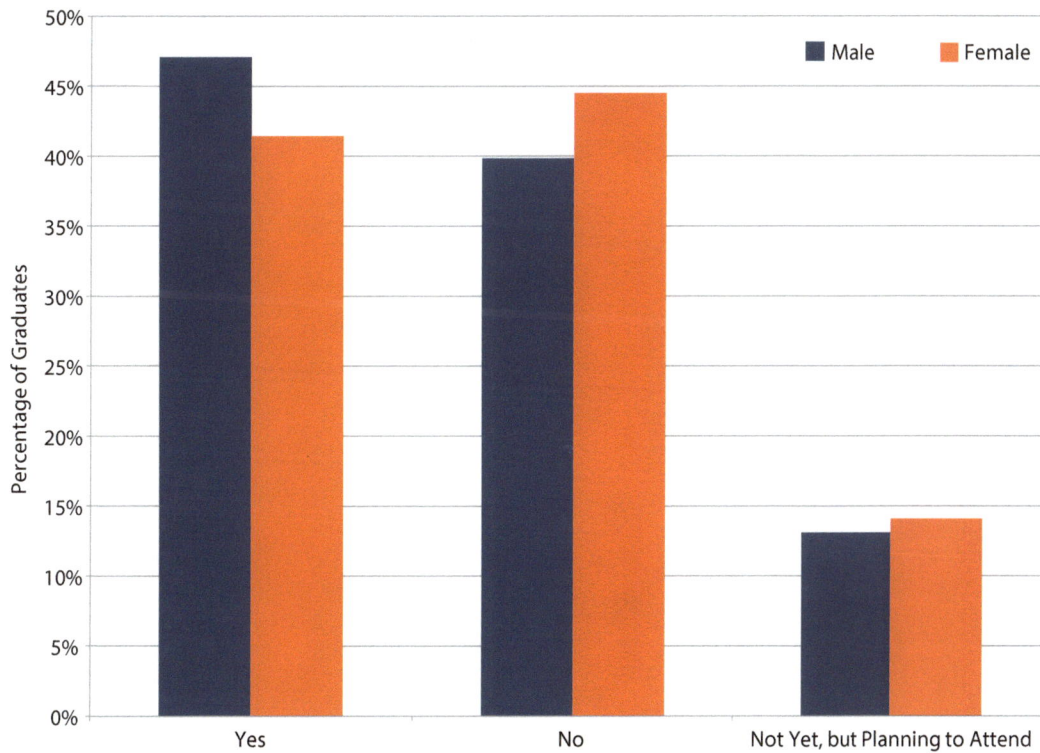

Number of field courses taken by graduates

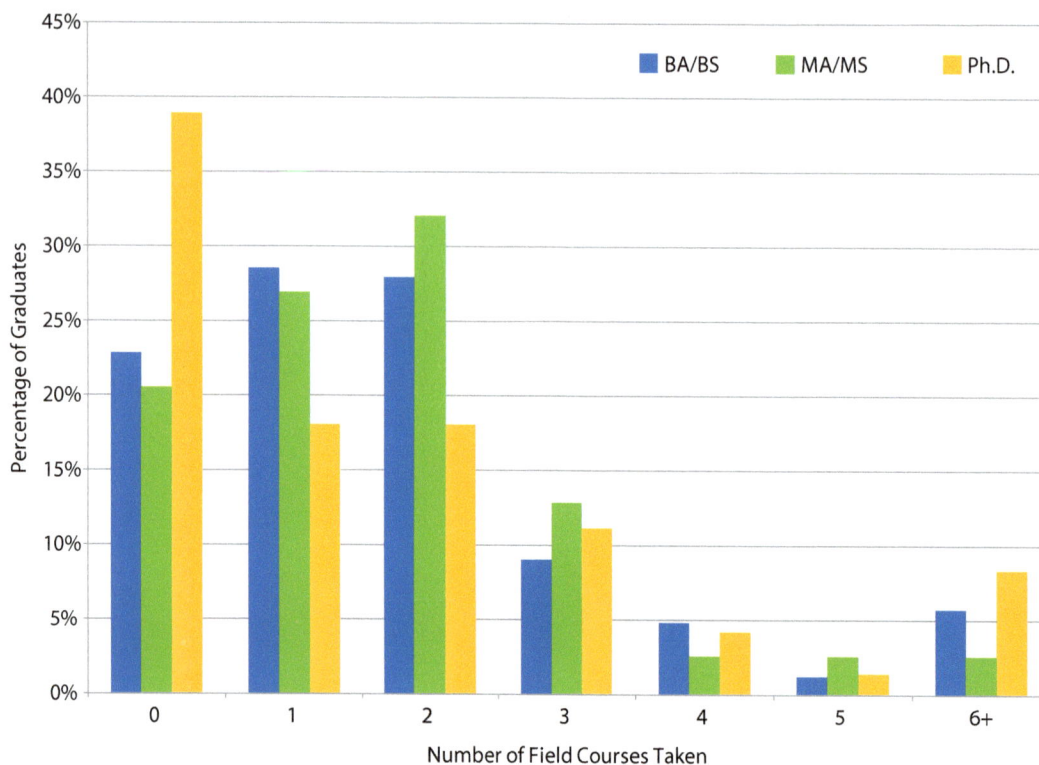

Courses taken with field experiences by graduates

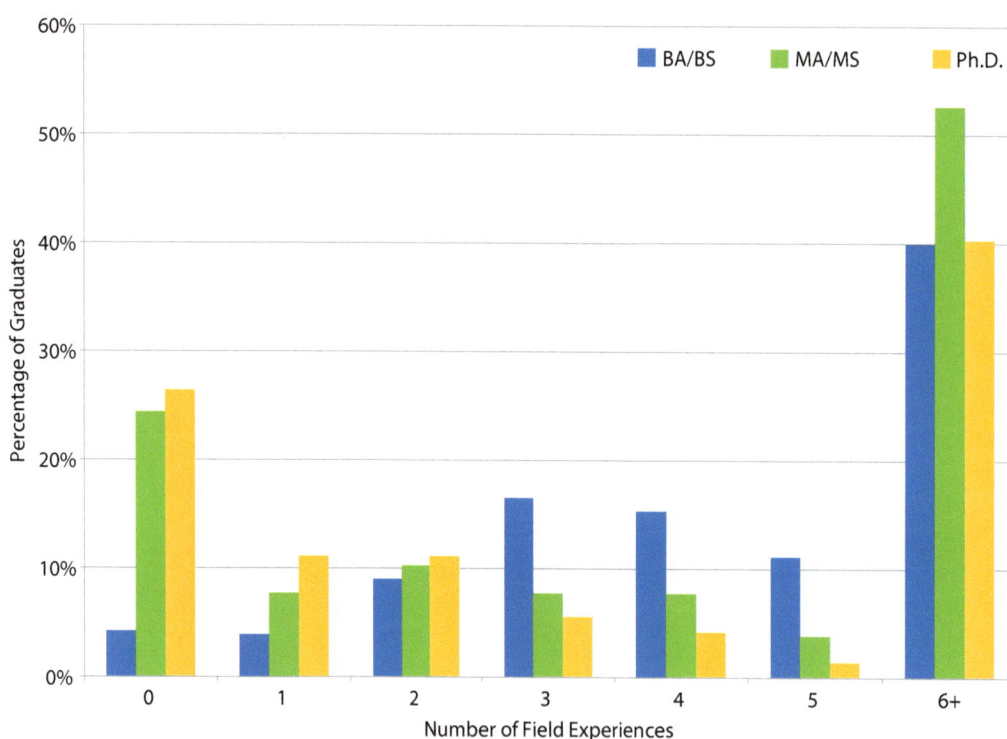

Image credit: Isidoros Kampolis, from AGI's 2016 Life as a Geoscientist contest

Cave survey in 'Kapareli' shaft, Boeotia, Central Greece. Checking the survey data on a survey sheet.

Research Experiences

The graduates were asked about their research experiences while working towards their degrees. If they indicated participation in at least one research experience, the graduates were then asked about their participation in faculty-directed research and self-directed research. If they indicated participation in self-directed research, they were asked to identify the basic research methodology used to conduct their research.

In 2016, there was a 9 percent increase in participation in research experiences by bachelor's graduates compared to 2015. However there was a 7 percent decrease in 2016 among master's graduates that did not participate in a research experience while working towards their degree compared to 2015.

Among graduate and undergraduate students that conducted self-directed research, the use of computer-based methods increase by 10 percent in 2016 compared to 2015, which is not surprising considering the increased emphasis on conducting research from large sets of earth science data. However, among the undergraduate students, 20 percent more male students indicated using computer-based methods for their research thank female students. This was the first year since 2014 that there was a gender difference in usages of research methods, and computer-based research methods continue to be the most used research method by graduate students conducting self-directed research. Future releases of AGI's Exit Survey data will indicate if this gender difference at the undergraduate level continues beyond 2016.

When asked about the importance of research experiences to the graduates' academic and professional development, 83 percent of bachelor's graduates, 84 percent of master's graduates and 99 percent of doctoral graduates rated these experiences as "very important." Outside of the classroom, these research experiences provide one of the best opportunities for students to utilize their critical thinking skills and learn how to work with uncertainty and imperfect data sets. It is encouraging to see high participation rates since the start of AGI's Exit Survey in research experiences at all degree levels.

Research methods utilized by graduates in their self-directed research

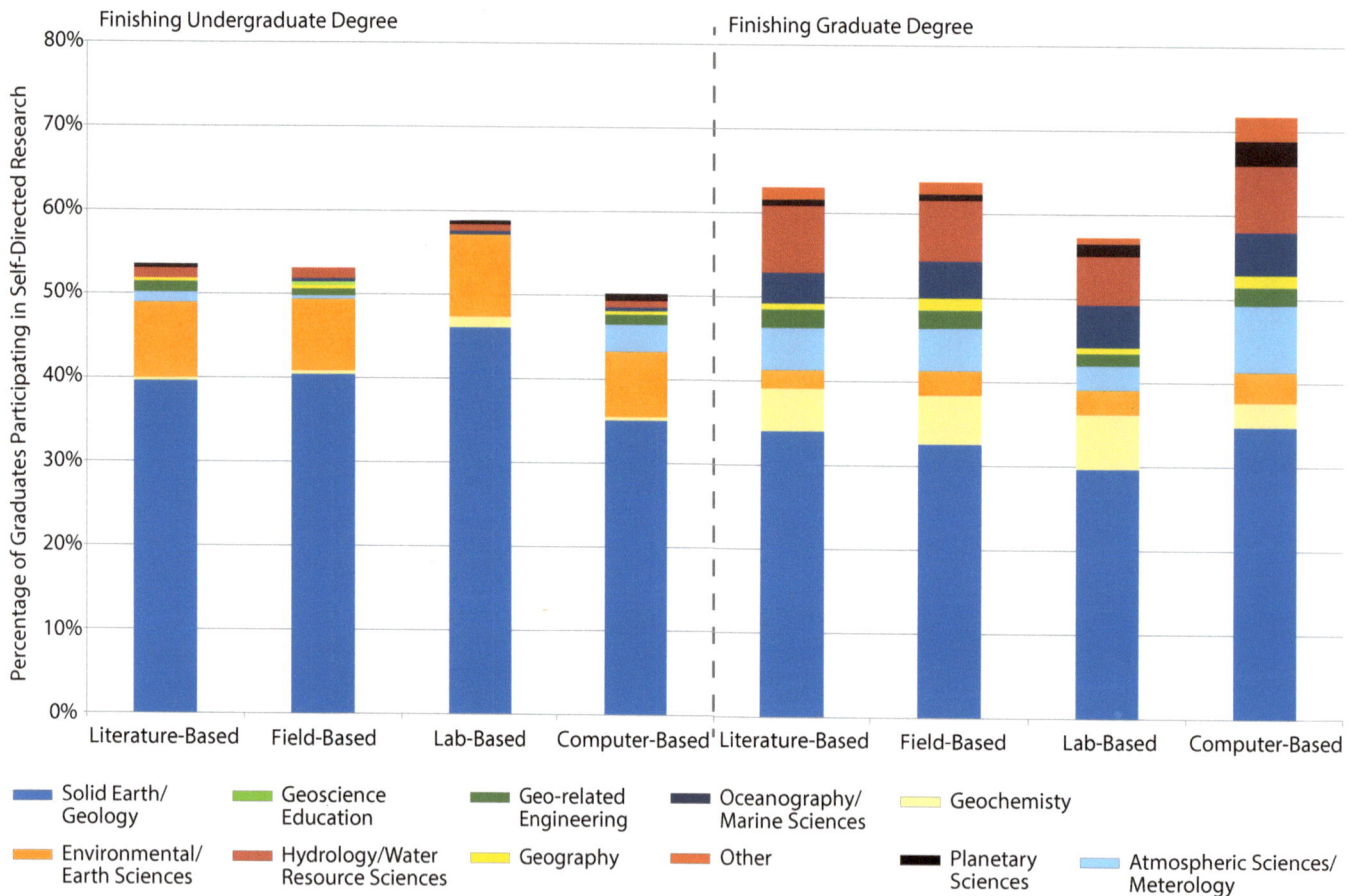

Participation rates of graduates in research experiences

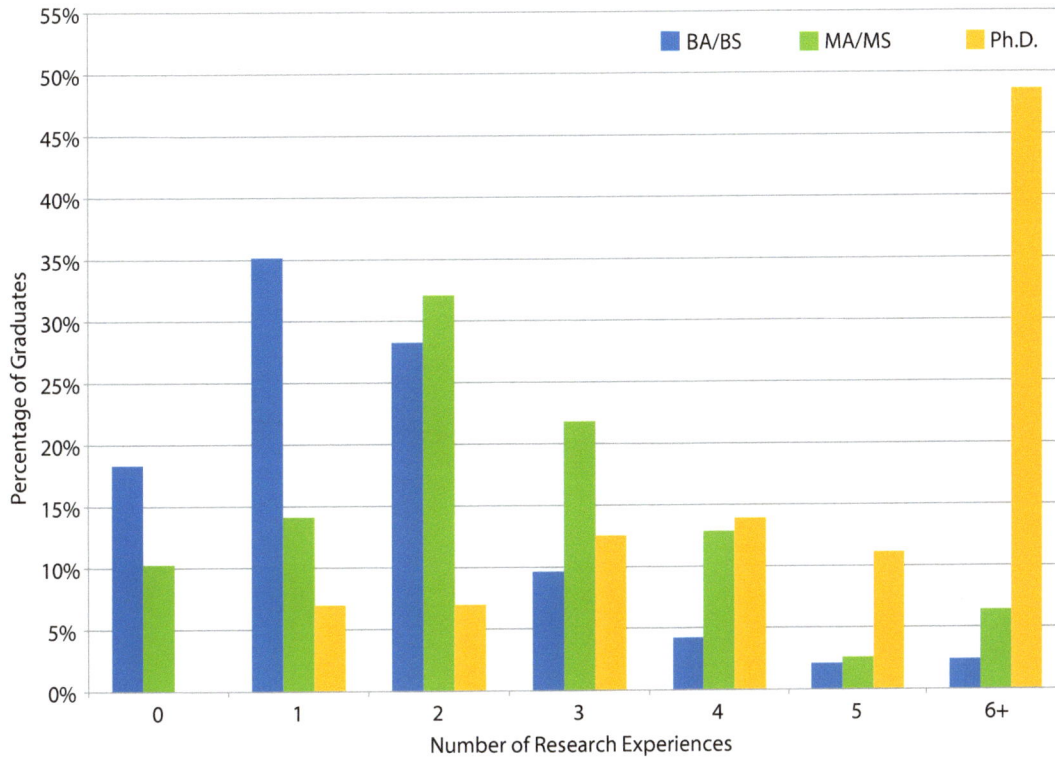

Student participation in faculty-directed and self-directed research

	BA/BS	MA/MS	Ph.D.
Faculty-Directed Research	58%	72%	90%
Self-Directed Research	74%	85%	100%

Research methods utilized by graduates in their self-directed research by gender

Finishing Undergraduate Degree

Finishing Graduate Degree

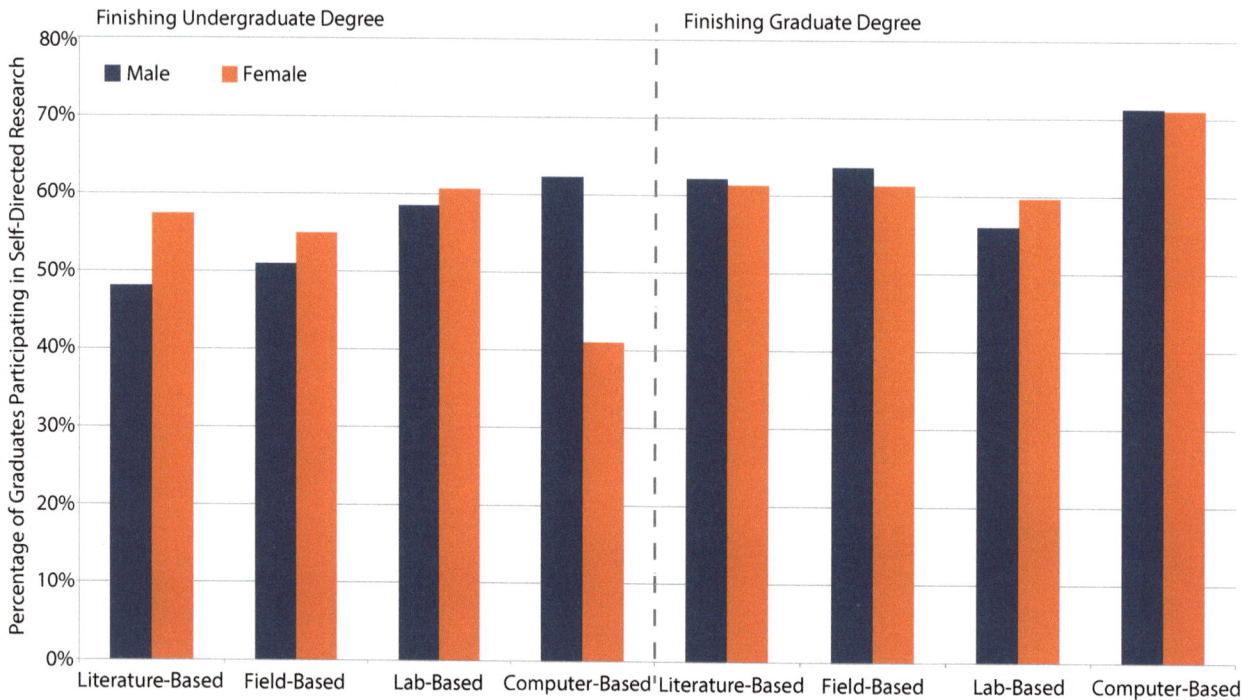

Percentage of Graduates Participating in Self-Directed Research

- Male
- Female

X-axis categories: Literature-Based, Field-Based, Lab-Based, Computer-Based (Finishing Undergraduate Degree); Literature-Based, Field-Based, Lab-Based, Computer-Based (Finishing Graduate Degree)

Student participation in research based on university classification**

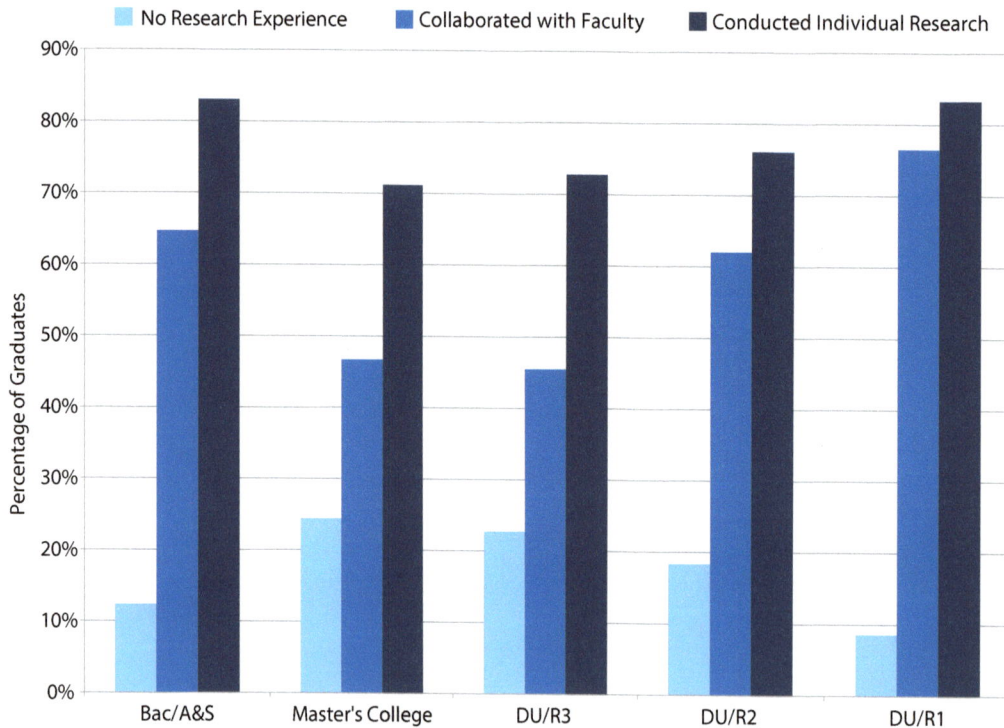

Legend: No Research Experience, Collaborated with Faculty, Conducted Individual Research

Percentage of Graduates

X-axis categories: Bac/A&S, Master's College, DU/R3, DU/R2, DU/R1

**See Appendix II for definitions of the Carnegie University Classification System

Image credit: Andres Carranco, from AGI's 2016 Life as a Geoscientist contest

Photo taken during the geological survey in the south of the Ecuadorian Andes, carried out by the National Institute of Metallurgical Mining Geological Research (INIGEMM), for generating the Geologic Map of Amaluza (scale 1:50.000) in 2015. Researchers and a local guide were walking down the Andes from the high moorland (3,200 m.a.s.l.) to the montane forest after a few days camping.

Future Plans: Working Toward a Graduate Degree

The graduates were asked if they have immediate plans to continue their education. Those indicating plans for a graduate degree after graduation were then asked to share the degree they would pursue and the field of interest for the degree.

From 2014-2016, the percentage of bachelor's graduates immediately planning to attend graduate school has ranged from 42 percent to 38 percent, and the percentage of master's graduates immediately planning to work towards another graduate degree has ranged from 20 percent to 27 percent. The majority of graduates plan to enter the workforce after graduation, whether that is in the geosciences or not. AGI's annual enrollment data updates have seen relatively stable enrollments in geoscience graduate programs, and conversations with departments have indicated that many of these graduate programs are at capacity, which has created competition for the open slots in these programs.

While the majority of bachelor's and master's graduates planning to attend graduate school were interested in a wide array of geoscience degree fields, there were a few recent graduates planning to obtain a graduate degree in a non-geoscience field, such as physics, biological sciences, law, math, and business. It's not clear if these recent graduates plan to use their geoscience knowledge in conjunction with these other degree fields or if these students are planning to move away from the geoscience workforce.

Students planning to attend graduate school after graduation

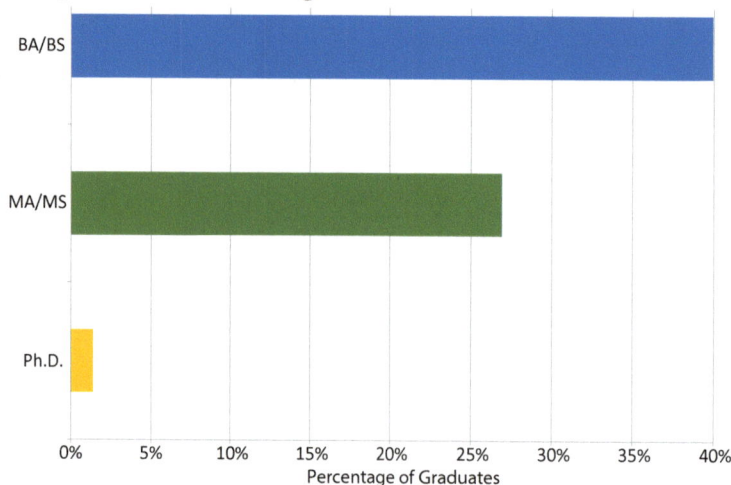

Students planning to attend graduate school after graduation by gender

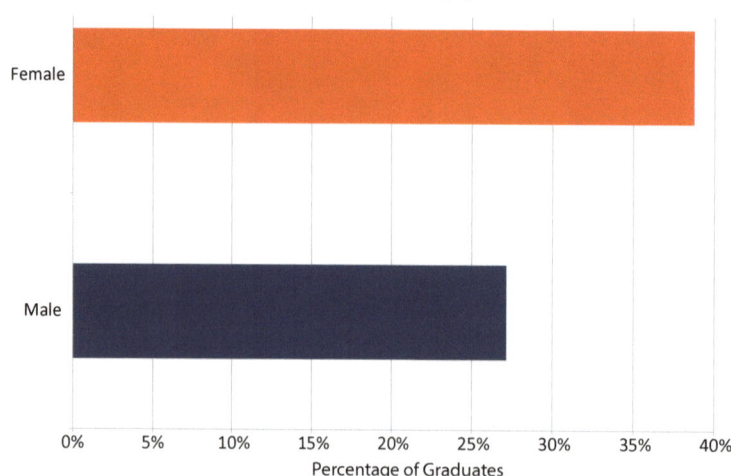

Students graduating with an undergraduate degree

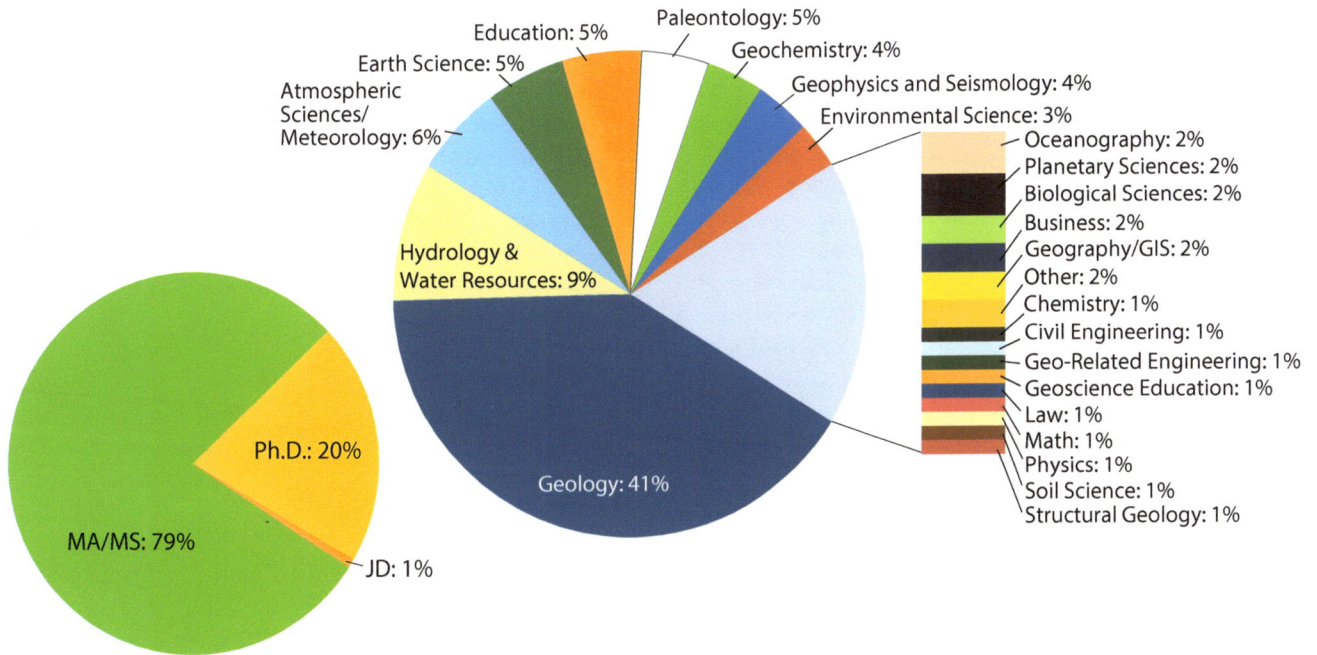

Possible Future Graduate Degree

- MA/MS: 79%
- Ph.D.: 20%
- JD: 1%

Possible Future Field of Study

- Geology: 41%
- Hydrology & Water Resources: 9%
- Atmospheric Sciences/Meteorology: 6%
- Earth Science: 5%
- Education: 5%
- Paleontology: 5%
- Geochemistry: 4%
- Geophysics and Seismology: 4%
- Environmental Science: 3%
- Oceanography: 2%
- Planetary Sciences: 2%
- Biological Sciences: 2%
- Business: 2%
- Geography/GIS: 2%
- Other: 2%
- Chemistry: 1%
- Civil Engineering: 1%
- Geo-Related Engineering: 1%
- Geoscience Education: 1%
- Law: 1%
- Math: 1%
- Physics: 1%
- Soil Science: 1%
- Structural Geology: 1%

Students graduating with a graduate degree

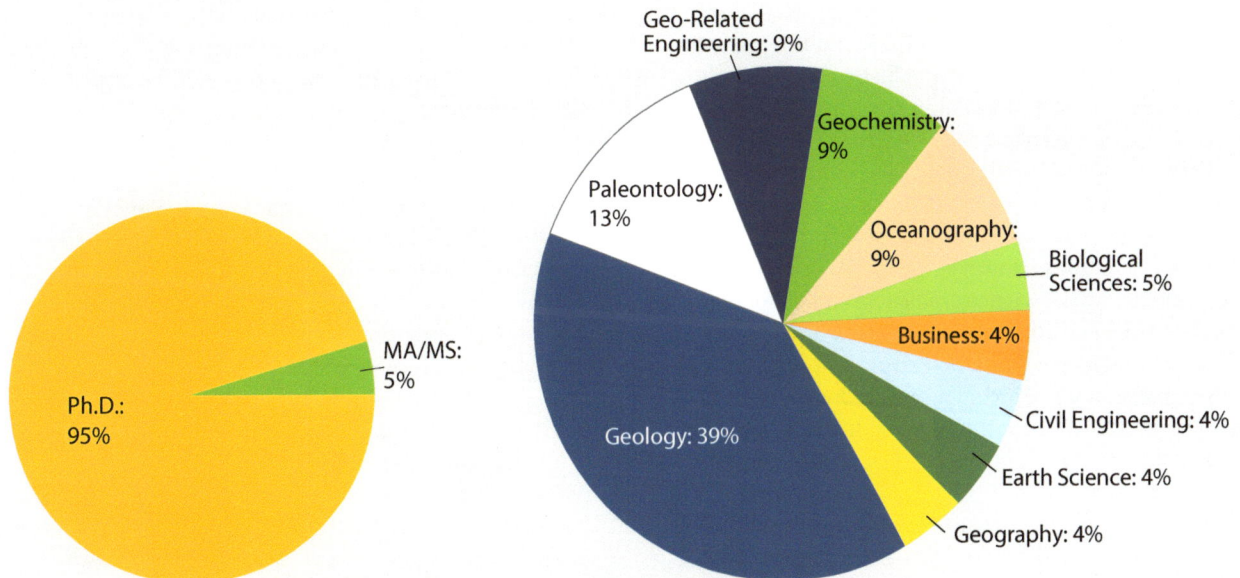

Possible Future Graduate Degree

- Ph.D.: 95%
- MA/MS: 5%

Possible Future Field of Study

- Geology: 39%
- Paleontology: 13%
- Geo-Related Engineering: 9%
- Geochemistry: 9%
- Oceanography: 9%
- Biological Sciences: 5%
- Business: 4%
- Civil Engineering: 4%
- Earth Science: 4%
- Geography: 4%

Future Plans: Working in the Geosciences

The graduates were asked if they had accepted or were seeking a job position within the geoscience workforce. If they had accepted a job, they were asked about these accepted positions. Because the graduates take the survey right around graduation, it is not surprising that there are still relatively high percentages of graduates at all degree levels still seeking employment. In 2016, 14 percent of bachelor's graduates, 32 percent of master's graduates, and 57 percent of doctoral graduates had secured a job in the geosciences at the time of graduation. Since 2014, doctoral graduates that found a geoscience job have dropped from 70 percent to 57 percent, and the percentage of master's graduates that found a geoscience was at its lowest in three years in 2016.

Among those graduates that were able to find a geoscience job, the bachelor's graduates tend to find positions in more industries than master's or doctoral graduates, and the jobs found by the master's and doctoral graduates tend to be in very traditional geoscience industries. The environmental services industry appears to be a very viable industry for bachelor's graduates with 31 percent of bachelor's graduates that found a job entering into that industry. The environmental services industry increased their hiring at the master's and doctoral degree levels as well. The oil and gas industry continues to conduct most of its hiring at the master's graduate level. Hiring of doctoral graduates by the oil and gas industry decreased by 10 percent in 2016 compared to 2015, and this year was the first year that no bachelor's graduates claimed finding a job within the oil and gas industry. There was also an increase in the hiring of graduates at all degree levels by the federal government in 2016. Since 2014 there have been changes in the employment dynamics in the oil and gas industry with an increased number of employees laid off from their positions. However, while the distribution of industries hiring geoscience graduates right out of school has changed since 2014, it may not be a direct cause of employment changes in the oil and gas industry. Oil and gas companies are still hiring recent graduates, but hiring may also be increasing within other industries and recent graduates may be looking for other opportunities.

As in previous years, the annual salaries for the geoscience jobs secured by the 2016 graduates show clear ranges by degree level, with bachelor's graduates earning $30,000-$60,000 and doctoral graduates earning $40,000-$80,000. Master's graduates tended to fall within two different ranges depending on the industry that hired them. The master's graduates that found jobs in environmental services and the federal government earned an annual salary from $40,000-$70,000, but master's graduates that found jobs in the oil and gas industry tended to earn an annual salary ranging from $80,000-$120,000.

Graduates that found geoscience employment were asked to identify the resources they used to find their job. Since 2014, graduates from all degree levels have noted the use of personal contacts as a major resource for finding their job. In 2016, approximately a quarter of bachelor's and doctoral graduates depended on faculty referrals to find their jobs, which supports the importance of the faculty in helping their students move forward after graduation. Beyond personal contacts, master's graduates have consistently relied on campus recruitment events for finding a job. The companies invited to these events are looking for master's students approaching graduation. Considering most master's graduates with a job at the time of graduation are working for the oil and gas industry, these campus recruitment events are likely dominated by energy companies or the energy companies in attendance are more interested in hiring immediately than the other industries. It also raises the question if the organizers of these events need to diversify the industry representation at the campus recruitment events.

The circular figure displays the connection between the degree fields of recent geoscience graduates from 2013-2016 (in color) to the industries where these geoscientists found their first job after graduation (in gray). The size of the bars along the outer edge of the circle represents the number of recent graduates that pursued a particular degree field and entered a particular industry. Each colored, inner ribbon connects a particular degree field with a job in a particular industry. The visualization shows the variety of industries available to graduates with a geoscience degree, as well as the complexity of the workforce and knowledge needed in the distinct industries.

Industries where graduating students have accepted a job within the geosciences

Graduates with a BA/BS

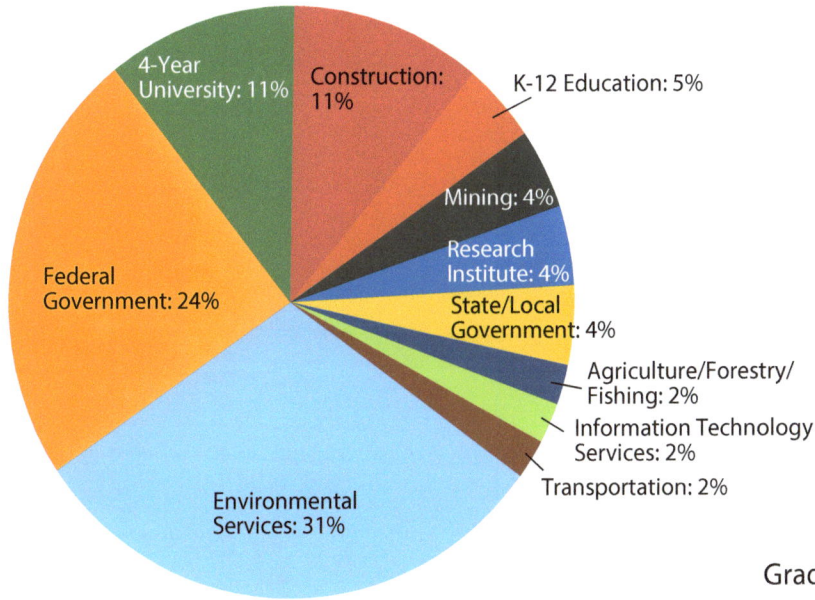

- 4-Year University: 11%
- Construction: 11%
- K-12 Education: 5%
- Mining: 4%
- Research Institute: 4%
- State/Local Government: 4%
- Agriculture/Forestry/Fishing: 2%
- Information Technology Services: 2%
- Transportation: 2%
- Environmental Services: 31%
- Federal Government: 24%

Graduates with a MA/MS

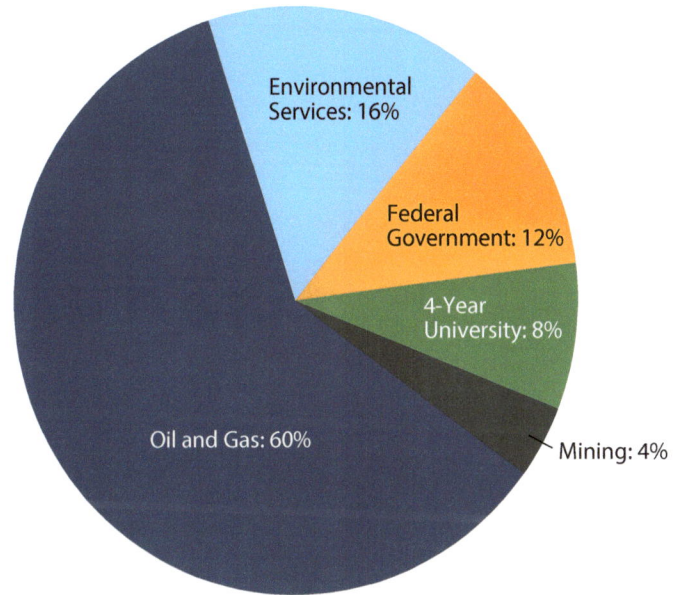

- Environmental Services: 16%
- Federal Government: 12%
- 4-Year University: 8%
- Mining: 4%
- Oil and Gas: 60%

Graduates with a Ph.D.

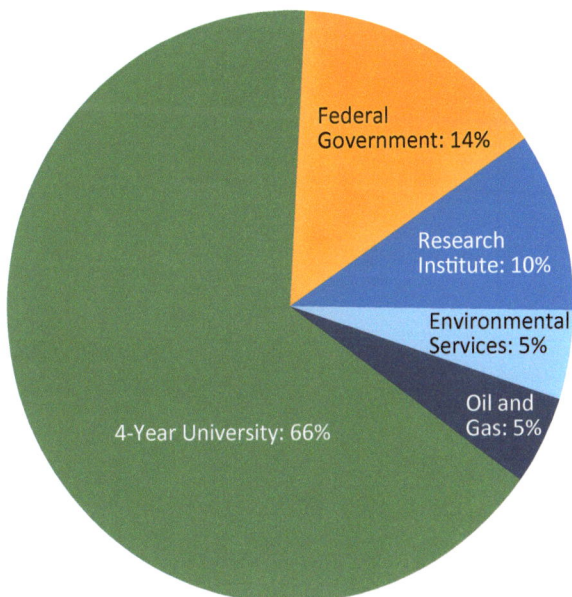

- Federal Government: 14%
- Research Institute: 10%
- Environmental Services: 5%
- Oil and Gas: 5%
- 4-Year University: 66%

Graduate students seeking or have accepted a position within the geosciences

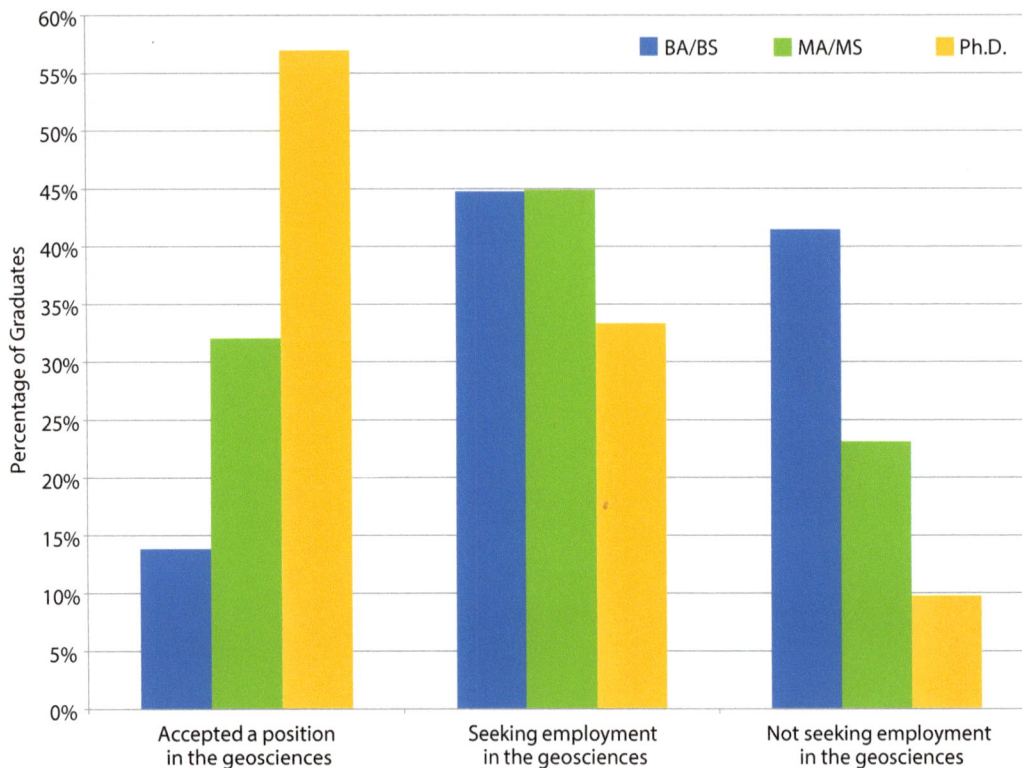

Graduate students seeking or have accepted a job within the geosciences by gender

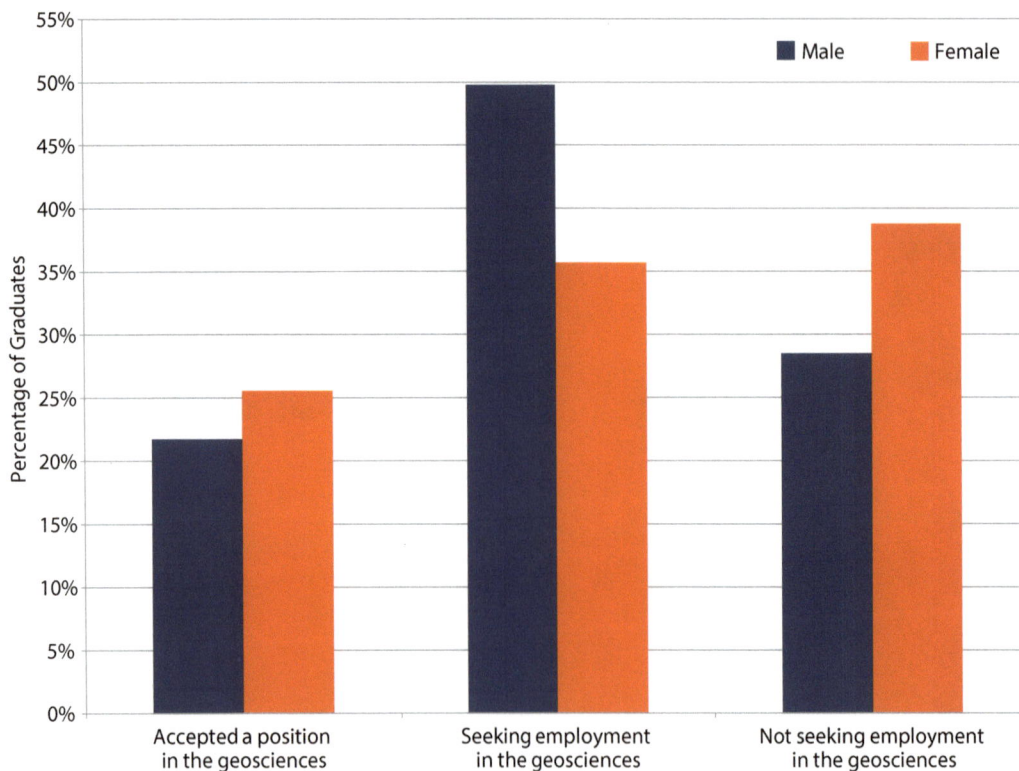

Starting salaries for graduates who accepted a job in the geosciences

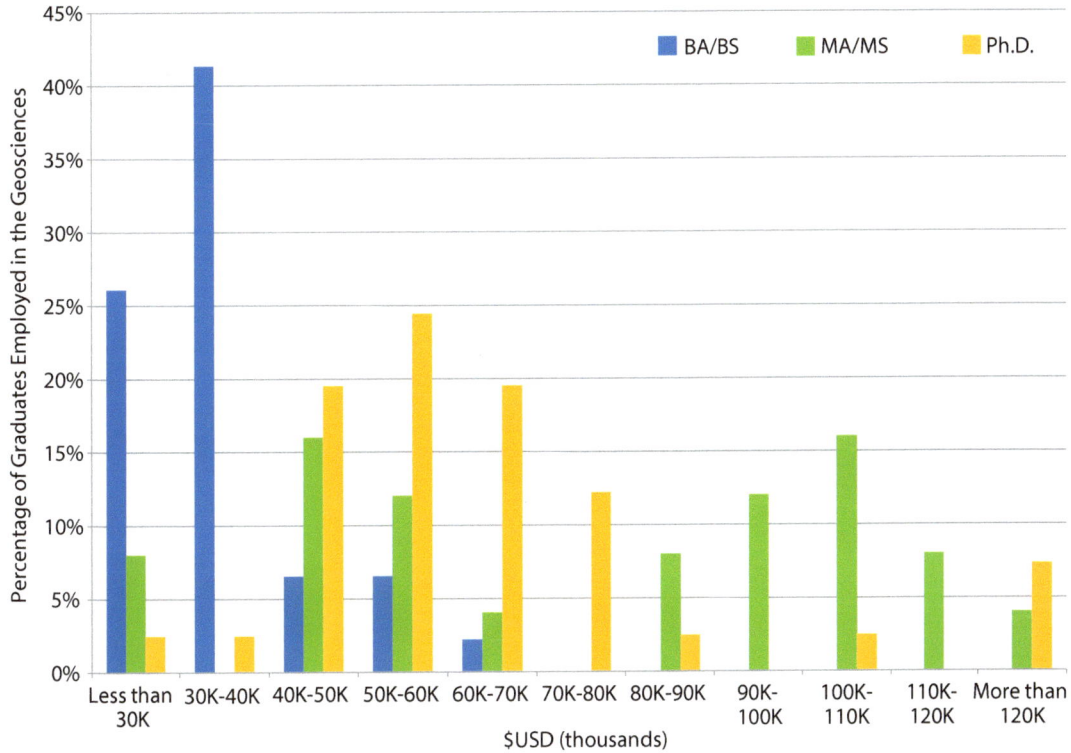

Additional compensation for graduates who accepted a job in the geosciences

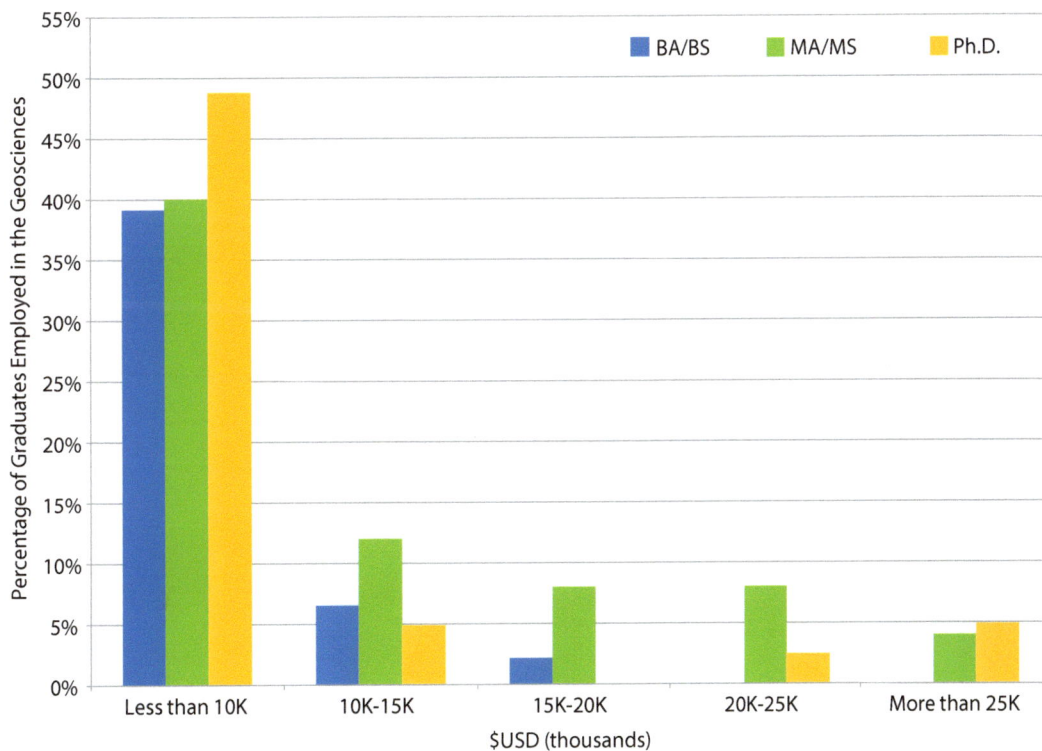

Resources identified by students as useful for finding geoscience jobs

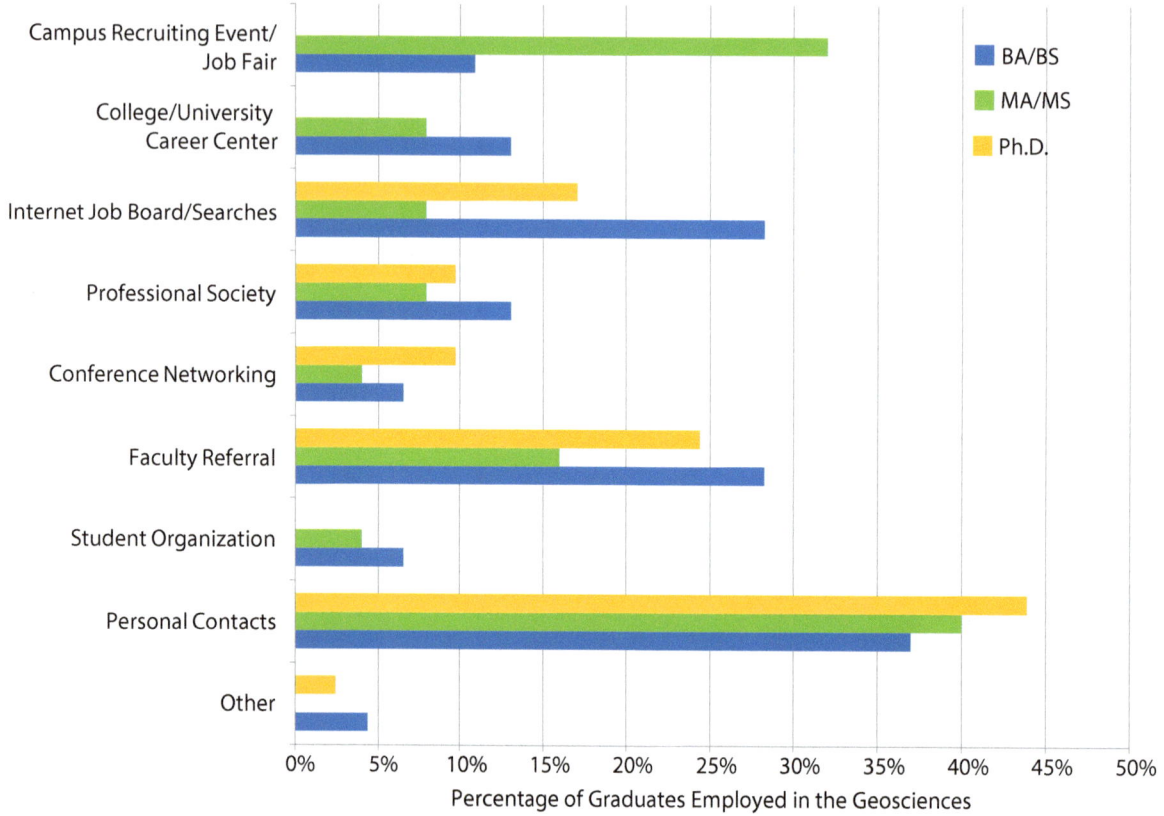

Other job opportunities offered to graduates who accepted a job in the geosciences

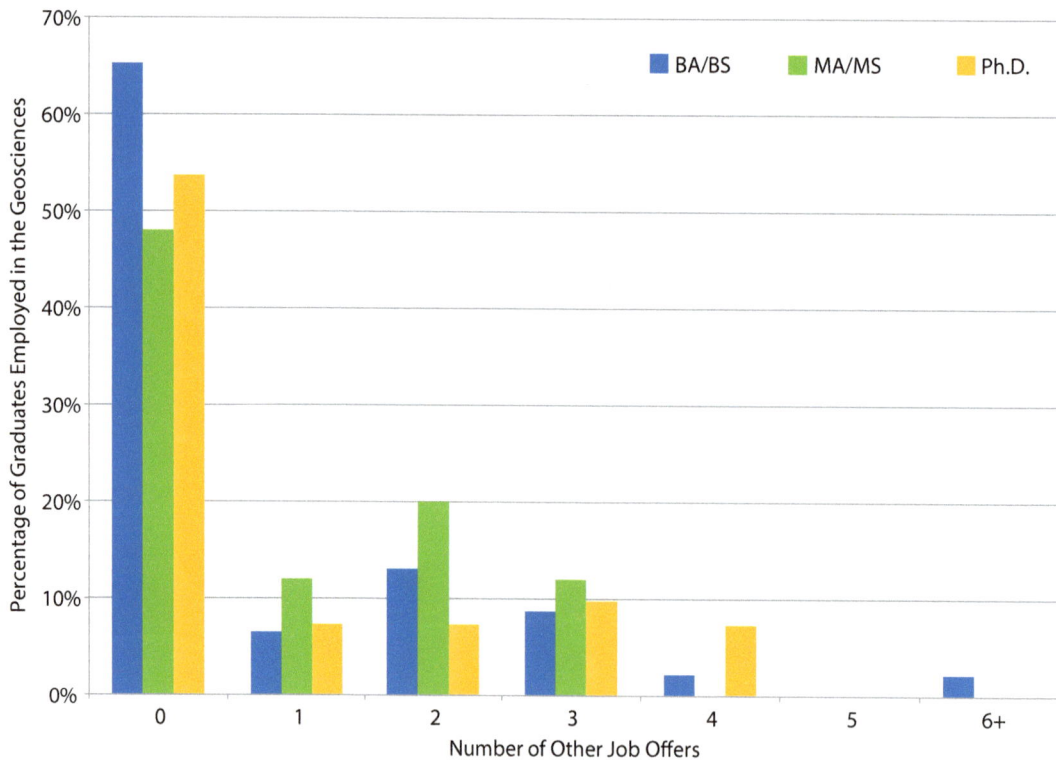

States where graduates found employment in the geosciences

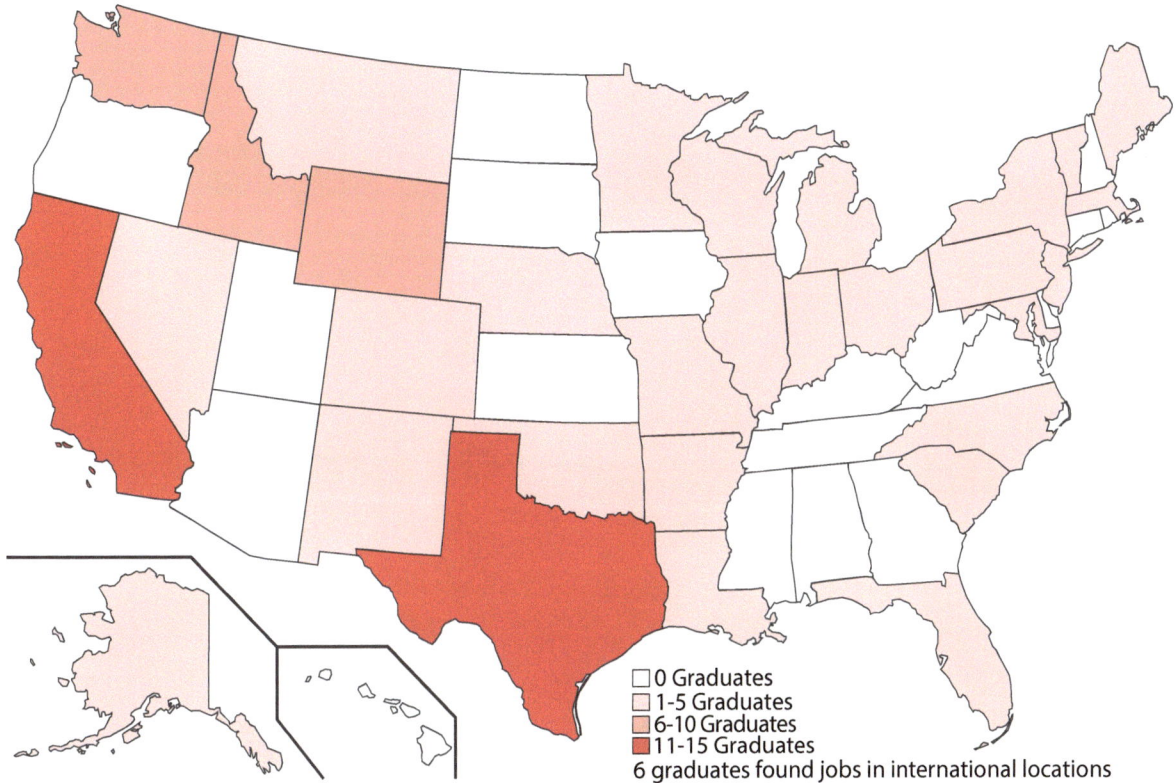

☐ 0 Graduates
☐ 1-5 Graduates
☐ 6-10 Graduates
☐ 11-15 Graduates

6 graduates found jobs in international locations

Industries of interest for graduating students seeking a job within the geosciences

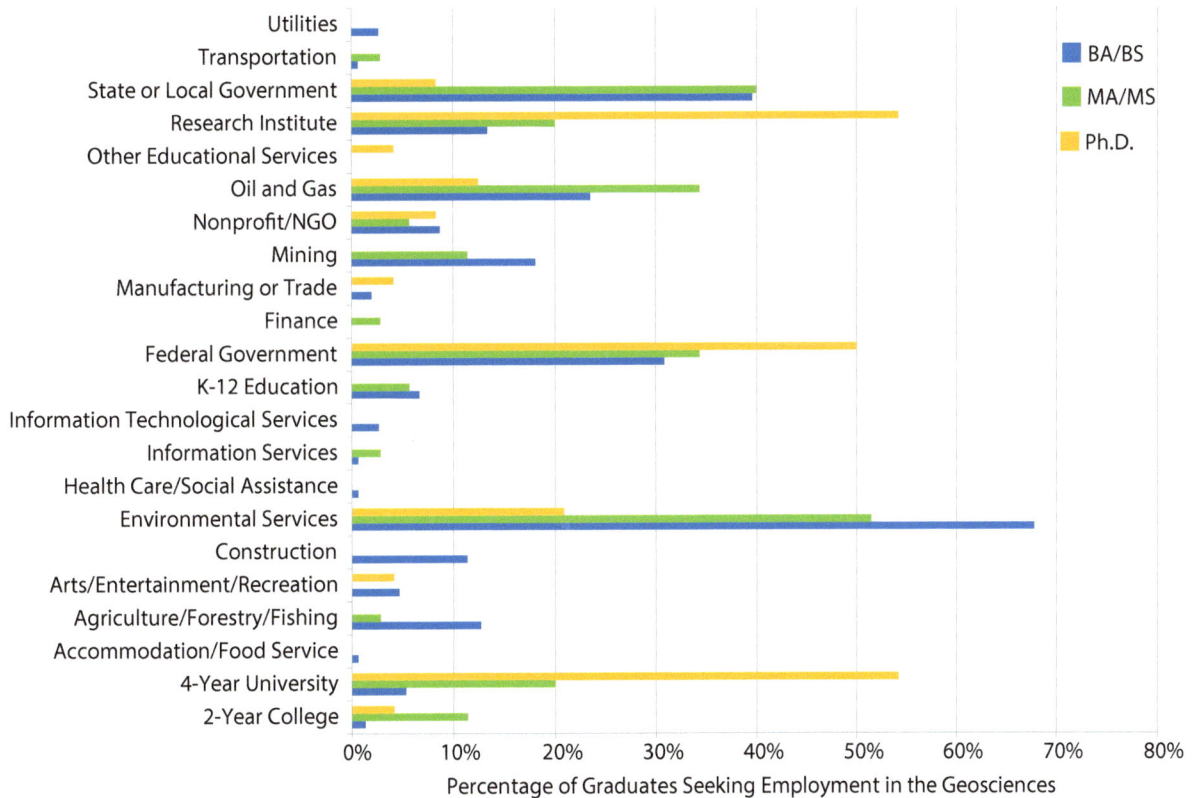

Legend: BA/BS, MA/MS, Ph.D.

Categories (top to bottom): Utilities, Transportation, State or Local Government, Research Institute, Other Educational Services, Oil and Gas, Nonprofit/NGO, Mining, Manufacturing or Trade, Finance, Federal Government, K-12 Education, Information Technological Services, Information Services, Health Care/Social Assistance, Environmental Services, Construction, Arts/Entertainment/Recreation, Agriculture/Forestry/Fishing, Accommodation/Food Service, 4-Year University, 2-Year College

Percentage of Graduates Seeking Employment in the Geosciences

Industries of geoscience graduates' first jobs by degree field for the past four years***

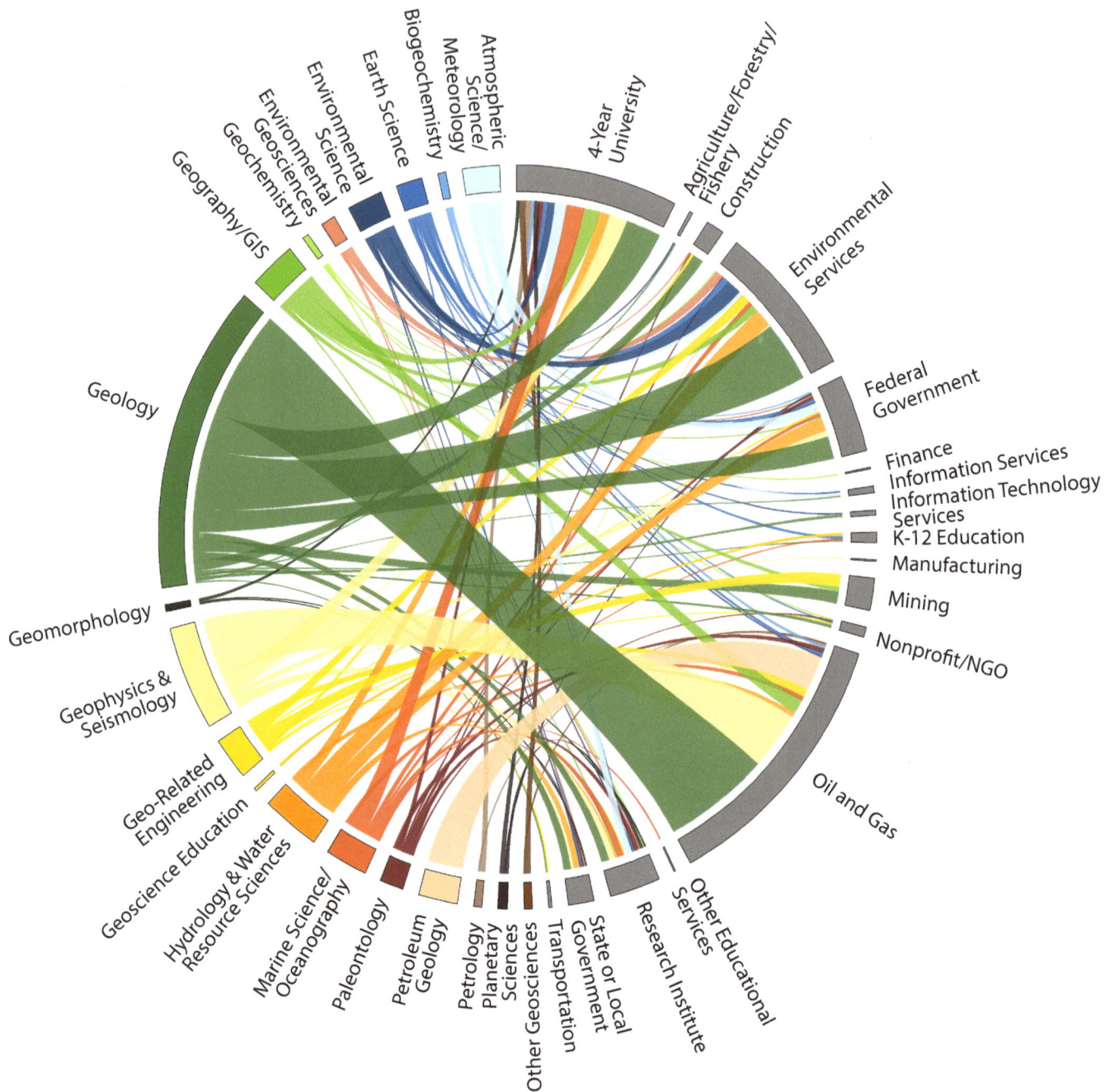

***The code for this visualization was modified from Kyzywinski, M. et al. Circos: an Information Aesthetic for Comparative Genomics.
Genome Res (2009) 19:1693–1645

Image credit: Justin Lawrence, from AGI's 2016 Life as a Geoscientist contest

McMurdo Sound, Antarctica. Drilling to measure sea ice thickness in front of McMurdo Base on Ross Island.

Future Plans: Working Outside of the Geosciences

Very few recent graduates are seeking or have secured jobs outside of the geosciences.

Each year, these graduates are asked why they pursued a job outside of the geosciences, and they respond with similar answers. Many graduates seek employment outside of the geosciences because they need a job immediately to help pay bills, were planning to enter the military, wanted to pursue other interests, or take time before going to graduate school. Often, graduates interested in K-12 education often do not equate teaching earth science as a geoscience job. However, AGI does consider them still within the geosciences community. In 2016, some different responses were provided from previous years related to the feelings of the recent economic downturn in the oil and gas industry affecting their job options. Multiple graduates thought jobs were not available to them in the oil and gas industry, either because they could not land an interview or because they felt they were competing for jobs with experienced individuals that had been fired from their previous oil and gas job. There were also some misconceptions about the working environments of different geoscience industries. Some of these issues and feelings of inadequacy in the geoscience workforce can be allayed among future graduates through internship-like experiences and more collaboration between geoscience departments and industry representatives.

While in 2016, there was a decrease in hiring in the oil and gas industry at the bachelor's and doctoral levels compared to recent years, the hiring presented in this report is at the time of graduation. The percentage distribution of industries hiring recent geoscience graduates would likely change if those questions were asked of recent graduates 6-12 months after graduation. It is important to provide realistic expectations and knowledge about their employment options while they are still in school.

Those graduates that had accepted a job outside of the geosciences were asked to provide more details about their positions. The industries hiring these students included K-12 education, non-profits, retail, military, and accommodation/food service. The majority of these graduates were offered starting annual salaries less than $30,000, and most of these jobs were found through personal contacts and internet searches.

Graduating students seeking or have accepted a job position outside the geosciences

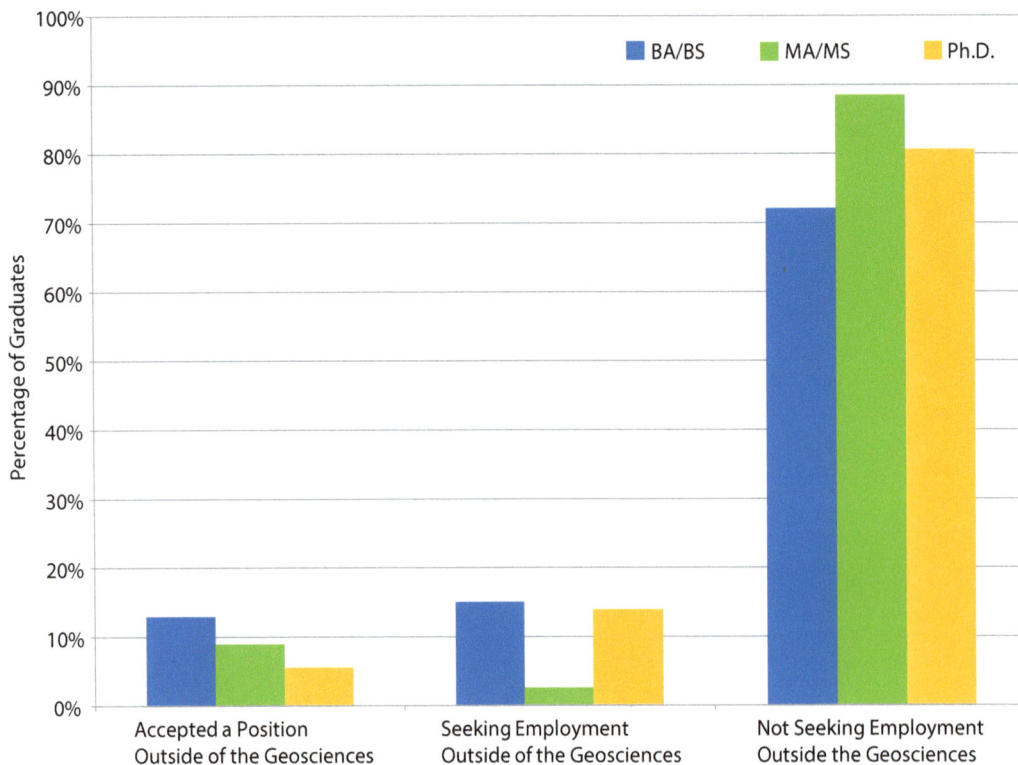

Industries where graduating students have accepted a job outside the geosciences

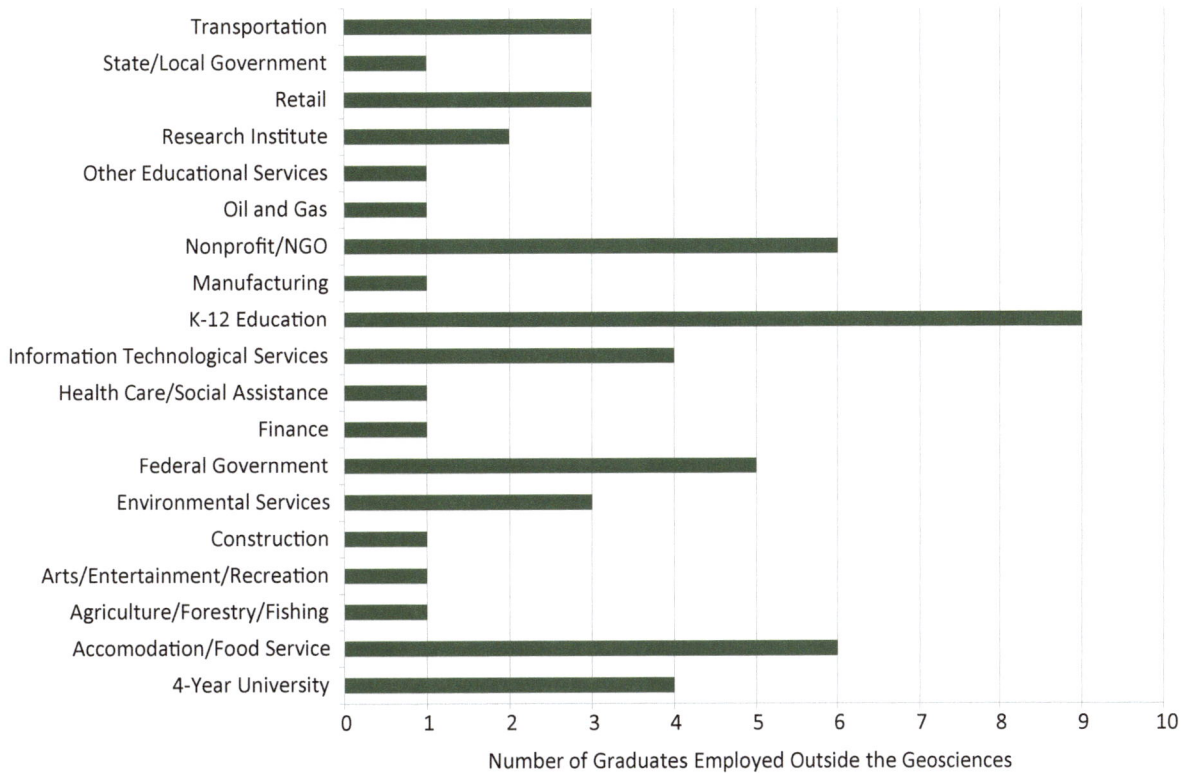

Number of Graduates Employed Outside the Geosciences

Starting salaries for graduating students that accepted a job outside the geosciences

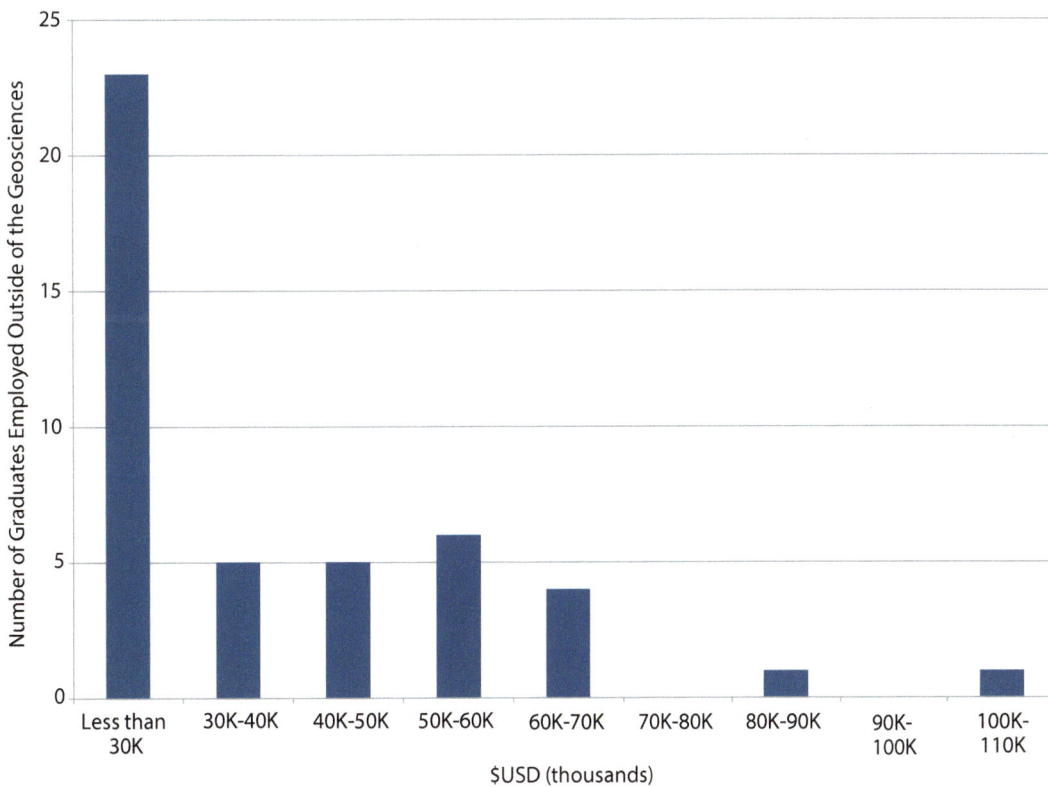

$USD (thousands)

Resources identified by graduating students as useful for finding non-geoscience jobs

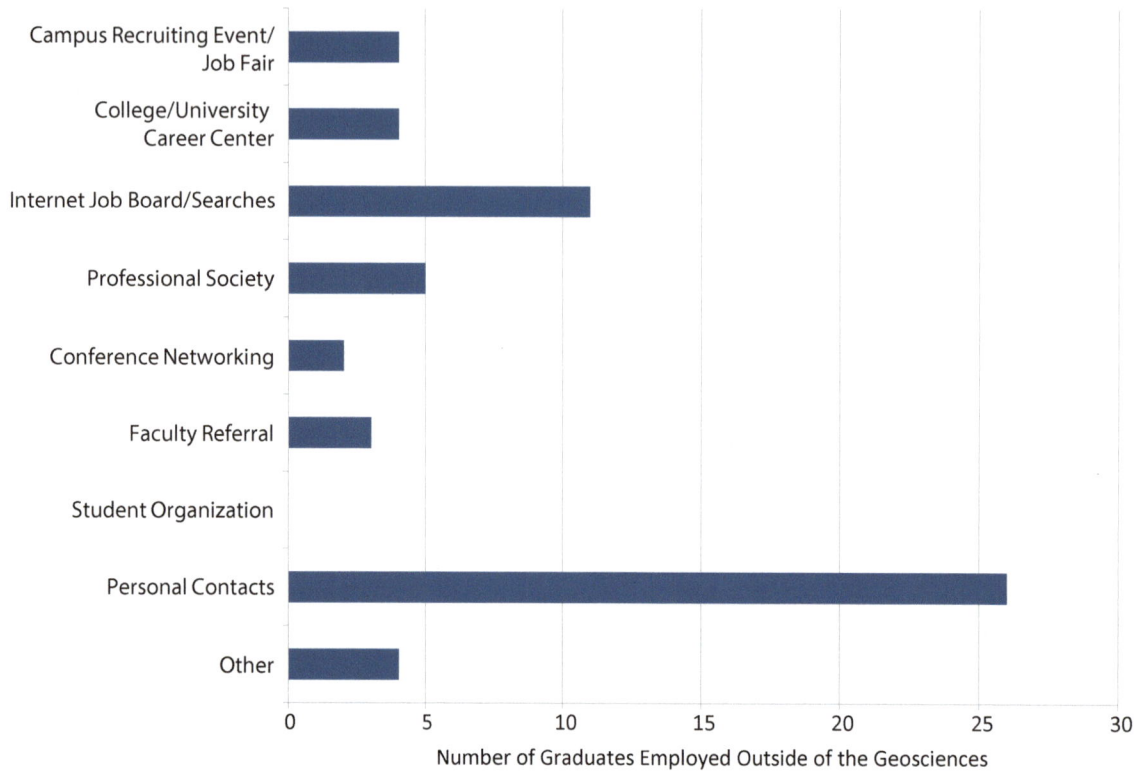

Number of Graduates Employed Outside of the Geosciences

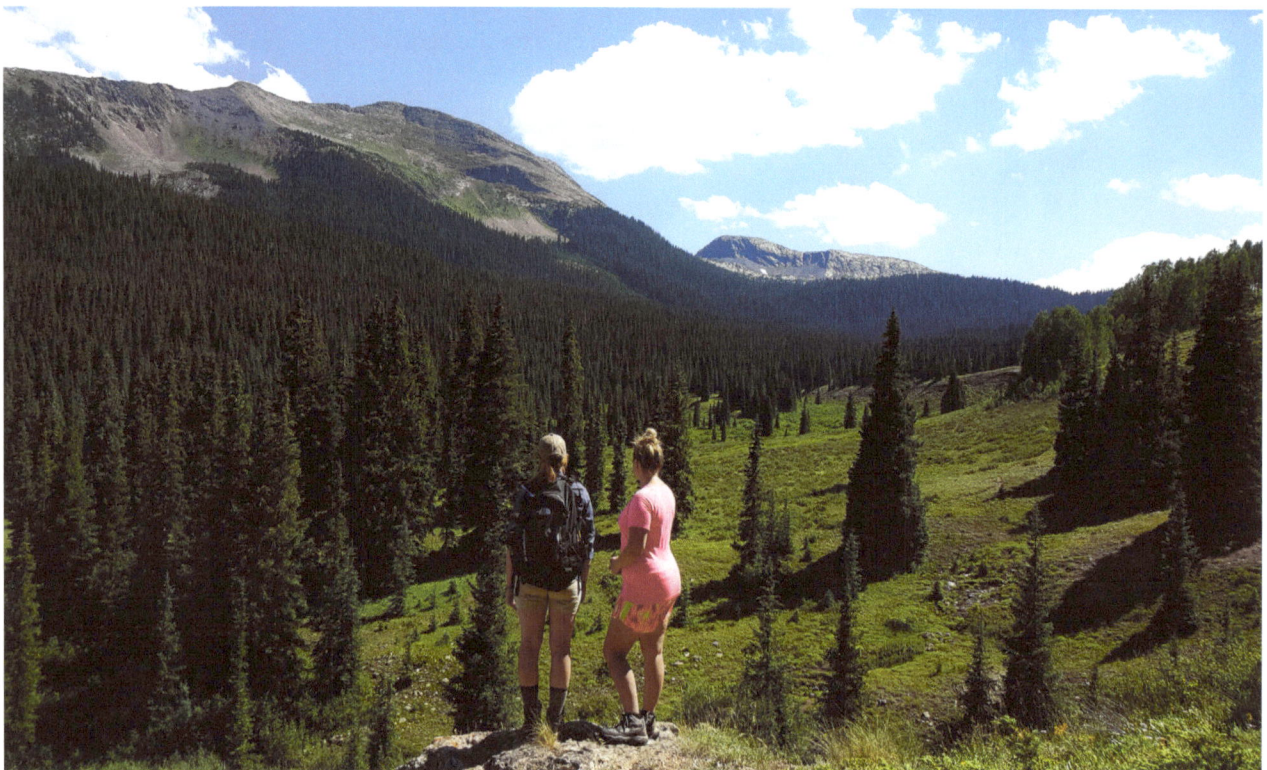

Image credit: Autumn Jones, from AGI's 2016 Life as a Geoscientist contest

Molas Lake, CO Field Camp II - University of Arkansas Little Rock "Women of Geosciences."

Image credit: Victoria Benson, from AGI's 2016 Life as a Geoscientist contest

In the midst of collecting suspended sediment and hydrochemical samples from the White River meltwater channel at Mount Rainier National Park, Washington. Early morning light illuminates the debris covered terminus of Emmons glacier in the background.

Pathway of Preparation for Entering the Geoscience Workforce

This figure is a Sankey diagram designed to show flow systems visually. In this case, the diagram displays the activities that help develop strong geoscience skills leading to the degree the graduates received and their immediate plans after graduation. The colored nodes represent the number of graduates participating in that experience and the gray ribbons represent the movement of individuals from one experience to the next. It sums up the geoscience experiences of the 2016 graduates, also shown through the series of graphs presented earlier in this report, to give an overall view of the pathway of preparation for the geoscience workforce among these graduates.

Moving forward, geoscience departments should strive to provide field and research experiences to all geoscience students through their programs, if they do not already, in order for effective development in critical geoscience skills and thinking. Future collaborations between universities,

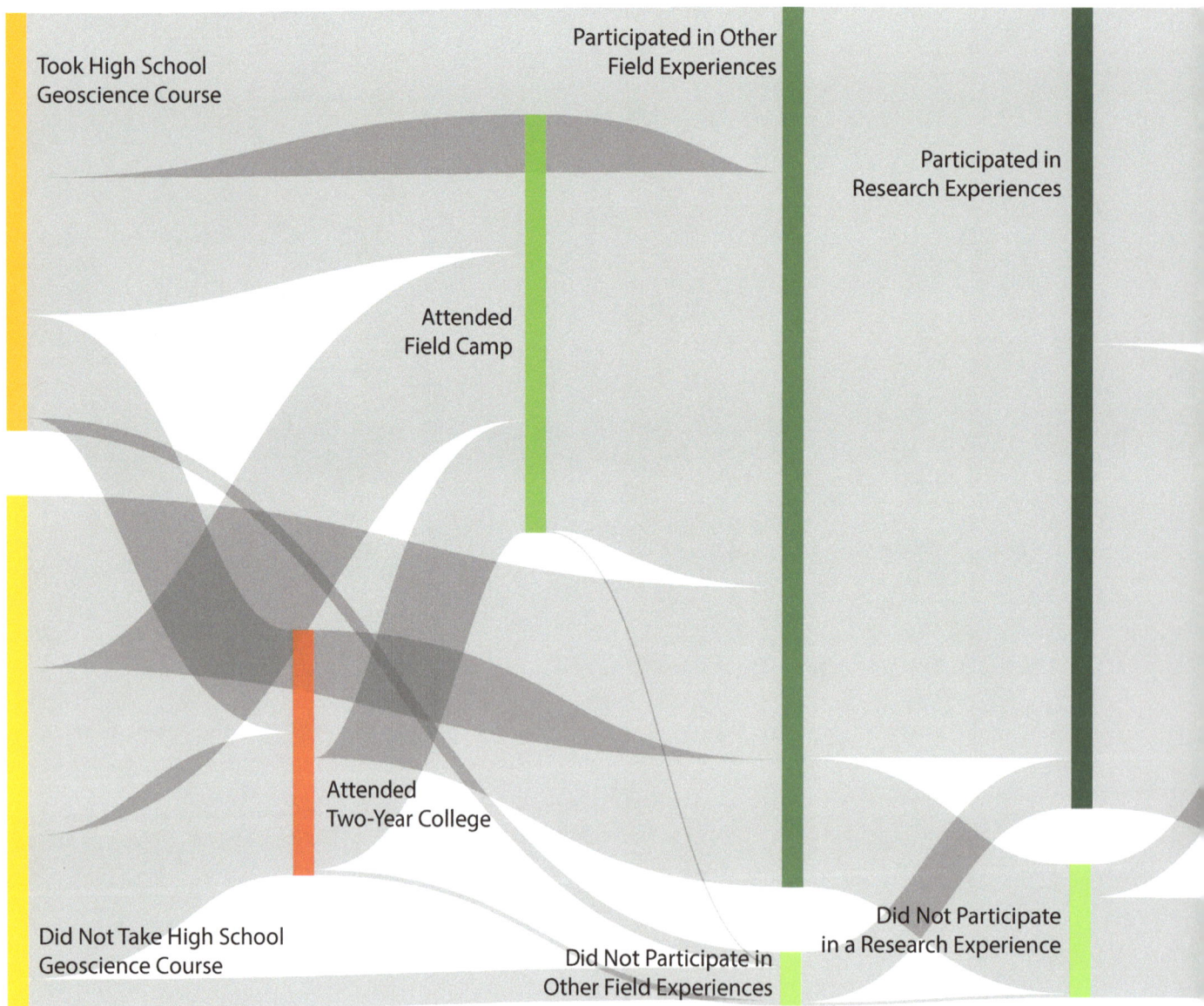

industries, and societies should begin to focus on developing internship-like experiences for more students in order to provide a more realistic understanding of the daily work within the various geoscience industries.

AGI's Geoscience Student Exit Survey will continue to collect data from geoscience graduates each year. Moving forward, AGI will reach out to former participants in the survey that are now in the workforce to see how their career pathways have developed. Variations of the survey are currently given in Canada and the UK with plans to expand to other countries in the future.

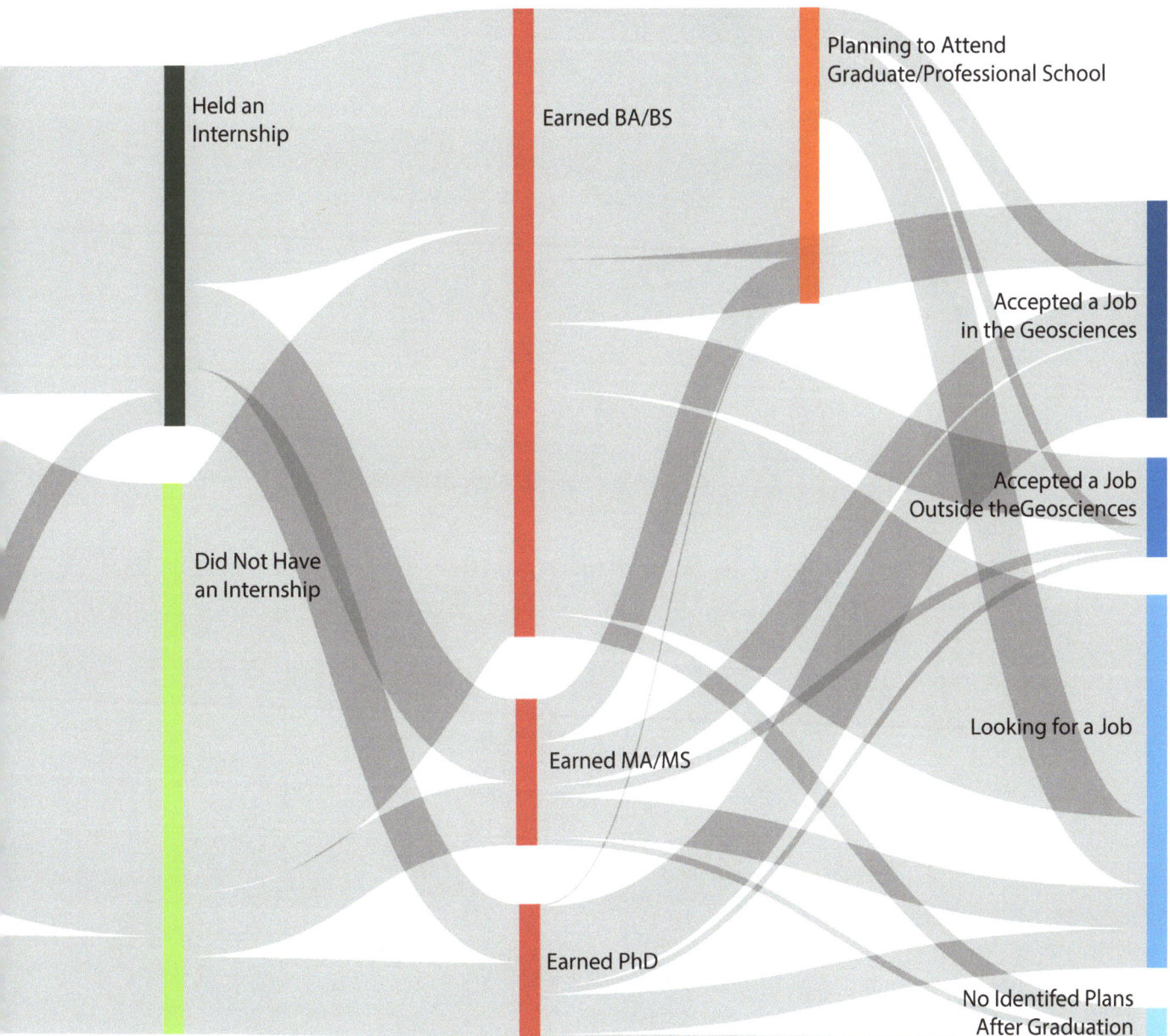

Appendices

Distribution of participating graduating students and departments*

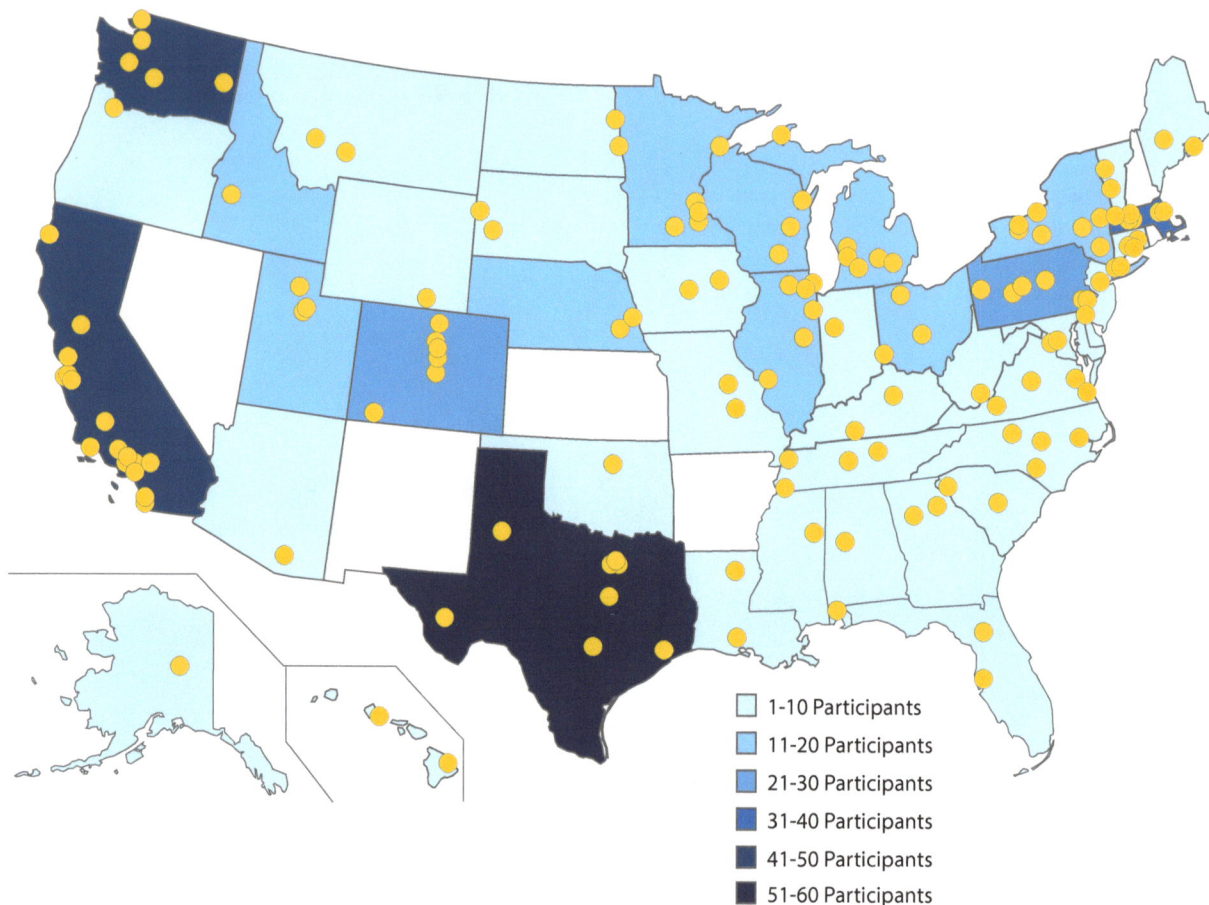

Legend:
- 1-10 Participants
- 11-20 Participants
- 21-30 Participants
- 31-40 Participants
- 41-50 Participants
- 51-60 Participants

Appendix I

The following is a list of all the institutions and departments with graduating students that took AGI's Geoscience Exit Survey in the 2015-2016 academic year.

University, Department

Amherst College, Department of Geology

Baylor University, Department of Geology

Black Hills State University, Department of Environmental Physical Sciences

Boise State University, Department of Geosciences

Bowling Green State University, Department of Geology

Brigham Young University-Idaho, Department of Geology

Bryn Mawr College, Department of Geology

Bucknell University, Department of Geology and Environmental Geosciences

California Institute of Technology, Department of Geological and Planetary Sciences

California State University-Bakersfield, Department of Geology

California State University-Monterey Bay, Moss Landing Marine Laboratories

California State University-Northridge, Department of Geological Sciences

Carleton College, Department of Geology

Castleton University, Department of Natural Sciences

Central Washington University, Department of Geological Sciences

Clemson University, Department of Environmental Engineering and Earth Sciences

Colby College, Department of Geology

College of the Atlantic, Department of Marine Sciences

College of Saint Rose, Department of Geology

College of William and Mary, Department of Geology

Colorado College, Department of Geology

Colorado School of Mines, Department of Geology and Geological Engineering

Colorado School of Mines, Department of Geophysics

Colorado State University, Department of Geosciences

Concord University, Department of Physical Sciences

Cornell University, Department of Earth and Atmospheric Sciences

CUNY City College, Department of Civil Engineering

CUNY Queens College, School of Earth and Environmental Sciences

East Carolina University, Department of Geological Sciences

Eastern Washington University, Department of Geology

Fort Lewis College, Department of Geosciences

George Mason University, Department of Atmospheric, Oceanic, and Earth Sciences

Georgia Institute of Technology, School of Earth and Atmospheric Sciences

Grand Valley State University, Department of Geology

Guilford College, Department of Geology and Earth Sciences

Gustavus Adolphus College, Department of Geology

Harvard University, Department of Earth and Planetary Sciences

Hope College, Department of Geological and Environmental Sciences

Humboldt State University, Department of Geology

Indiana University of Pennsylvania, Department of Geoscience

Iowa State University, Department of Geological and Atmospheric Sciences

Macalester College, Department of Geology

Miami University of Ohio, Department of Geology and Environmental Earth Science

Michigan State University, Department of Geological Sciences

Michigan Technological University, Department of Geological/Mining Engineering and Sciences

Middle Tennessee State University, Department of Geosciences

Mississippi State University, Department of Geosciences

Missouri University of Science and Technology, Department of Geosciences and Geological and Petroleum Engineering

Montana State University, Department of Geology

Montana State University, Department of Physics

Montana Tech of the University of Montana, Department of Geophysical Engineering

North Carolina State University, Department of Marine, Earth, and Atmospheric Sciences

North Dakota State University, Department of Geosciences

Northern Illinois University, Department of Geology and Environmental Geosciences

Northwestern University, Department of Earth and Planetary Sciences

Ohio State University, Department of Civil Environmental and Geodetic Engineering

Ohio State University, School of Earth Sciences

Oklahoma State University, Department of Geology

Old Dominion University, Department of Ocean, Earth, and Atmospheric Sciences

Olivet Nazarene University, Department of Geological Sciences

Pacific Lutheran University, Department of Geoscience

Pennsylvania State University, Department of Geosciences

Pomona College, Department of Geology

Purdue University, Department of Biological Sciences

Purdue University, Department of Earth and Atmospheric Sciences

Reed College, Department of Chemistry

Rutgers University, Department of Earth and Planetary Sciences

San Diego State University, Department of Geological Sciences

Slippery Rock University, Department of Geography, Geology, and the Environment

Smith College, Department of Geosciences

South Dakota School of Mines and Technology, Department of Geology and Geological Engineering

Southern Methodist University, Department of Earth Sciences

St. Louis University, Department of Earth and Atmospheric Sciences

St. Norbert College, Department of Geology

Stanford University, Department of Earth System Science

Stanford University, Department of Geological Sciences

Stanford University, Department of Geophysics

Sul Ross University, Department of Biology, Geology, and Physical Sciences

SUNY Geneseo, Department of Geological Sciences

SUNY New Paltz, Department of Physics and Astronomy

SUNY Oneonta, Department of Earth and Atmospheric Sciences

SUNY Oswego, Department of Atmospheric and Geological Sciences

Temple University, Department of Earth and Environmental Science

Tennessee Tech University, Department of Earth Sciences

Texas Tech University, Department of Geosciences

Tufts University, Department of Earth and Ocean Sciences

University of Alabama, Department of Geological Sciences

University of Alaska-Fairbanks, Department of Geosciences

University of Arizona, Department of Hydrology

University of California-Berkeley, Department of Earth and Planetary Science

University of California-Davis, Department of Earth and Planetary Sciences

University of California-Los Angeles, Department of Earth, Planetary and Space Sciences

University of California-Riverside, Department of Earth Sciences

University of California-San Diego, Scripps Institution of Oceanography

University of California-Santa Barbara, Department of Earth Science

University of California-Santa Cruz, Department of Earth and Planetary Sciences

University of Colorado at Boulder, Department of Geological Sciences

University of Colorado at Boulder, Department of Allied Mathematics

University of Colorado at Boulder, Department of Aerospace Engineering Sciences

University of Colorado Denver, Department of Geography and Environmental Sciences

University of Connecticut, Department of Marine Science

University of Delaware, Department of Geological Sciences

University of Delaware, Department of Plant and Soil Sciences

University of Florida, Department of Agricultural and Biological Engineering

University of Florida, Department of Geological Sciences

University of Georgia, Department of Geography

University of Georgia, Department of Marine Sciences

University of Hawaii-Hilo, Department of Geology

University of Hawaii-Manoa, School of Ocean & Earth Science & Technology

University of Houston, Department of Earth and Atmospheric Sciences

University of Illinois, Department of Geology

University of Kentucky, Department of Earth and Environmental Sciences

University of Louisiana at Lafayette, Department of Geology

University of Louisiana at Monroe, Department of Atmospheric Science

University of Maryland, Department of Atmospheric and Oceanic Sciences

University of Maryland, Department of Geology

University of Massachusetts, Department of Geosciences

University of Memphis, Department of Earth Sciences

University of Michigan, Department of Earth and Environmental Sciences

University of Minnesota, Department of Earth Sciences

University of Minnesota, Department of Physics

University of Minnesota-Duluth, Department of Geosciences

University of Missouri, Department of Geological Sciences

University of Nebraska-Lincoln, Department of Earth and Atmospheric Sciences

University of Nebraska-Omaha, Department of Geography/Geology

University of North Carolina at Pembroke, Department of Geology and Geography

University of North Dakota, Department of Earth System Science and Policy

University of Northern Iowa, Department of Earth Sciences

University of Rochester, Department of Physics and Astronomy

University of South Alabama, Department of Earth Sciences

University of South Carolina, Department of Earth and Ocean Sciences

University of South Florida, School of Geosciences

University of Tennessee at Martin, Department of Agriculture and Applied Sciences

University of Texas at Arlington, Department of Earth and Environmental Sciences

University of Texas at Austin, Jackson School of Geosciences

University of Texas at Dallas, Department of Geosciences

University of Utah, College of Mines and Earth Sciences

University of Vermont, Department of Geology

University of Virginia, Department of Environmental Sciences

University of Washington, Department of Atmospheric Sciences

University of Washington, Department of Civil and Environmental Engineering

University of Washington, Department of Earth and Space Sciences

University of Washington, Department of Oceanography

University of Wisconsin-Madison, Department of Geology and Geophysics

University of Wisconsin-Oshkosh, Department of Geology

University of Wyoming, Department of Geology and Geophysics

Virginia Polytechnic Institute and State University, Department of Geosciences

Weber State University, Department of Geosciences

Wesleyan University, Department of Earth and Environmental Sciences

Western Kentucky University, Department of Geography and Geology

Western Michigan University, Department of Geosciences

Western Washington University, Department of Geology

Westminster College, Department of Biology and Environmental Sciences

Wheaton College, Department of Geology and Environmental Science

Williams College, Department of Geosciences

Yale University, Department of Geology and Geophysics

Appendix II

Carnegie Classifications of Institutions of Higher Learning
(http://carnegieclassifications.iu.edu//resources/links.php)

This classification system was used for some of the analysis of the Spring 2013 results of AGI's Geoscience Student Exit Survey. The following are the definitions for the classification system and the participating institutions belonging to each category as defined and categorized by the Carnegie Foundation for the Advancement of Teaching.

Baccalaureate Colleges — Arts & Sciences (Bac/A&S)

Baccalaureate Colleges — Diverse Fields (Bac/Diverse)

Includes institutions where baccalaureate degrees represent at least 50 percent of all degrees but where fewer than 50 master's degrees or 20 doctoral degrees were awarded during the update year. (Some institutions above the master's degree threshold are also included). Excludes Special Focus Institutions and Tribal Colleges.

Institutions in which at least half of the bachelor's degree majors in arts and sciences fields were included in the "Arts and Sciences" group, while the remaining institutions were included in the "Diverse Fields" group. For more information about the identification of baccalaureate colleges, please visit the description of the Basic Classification Methodology (http://carnegieclassifications.iu.edu/methodology/basic.php).

Exit Survey Departments (Bac/A&S):
Amherst College
Bryn Mawr College
Bucknell University
Carleton College
Colby College
College of the Atlantic
Colorado College
Fort Lewis College
Guilford College
Gustavus Adolphus College
Hope College
Macalester College
Pomona College

Reed College
Smith College
St. Norbert College
Westminster College
Wheaton College
Williams College

Exit Survey Departments (Bac/Diverse)*
Brigham Young University-Idaho
Castleton University
Montana Tech of the University of Montana

Master's Colleges and Universities — Larger Programs (Master's/L)

Master's Colleges and Universities — Medium Programs (Master's/M)

Master's Colleges and Universities — Smaller Programs (Master's/S)

Generally includes institutions that awarded at least 50 master's degrees and fewer than 20 doctoral degrees during the update year (with occasional exceptions). Excludes Special Focus Institutions and Tribal Colleges.

For more information about the determination of the size of master's programs, please visit the description of the Basic Classification Methodology (http://carnegieclassifications.iu.edu/methodology/basic.php).

Exit Survey Departments (Master's/L):
California State University-Bakersfield
California State University-Northridge
Central Washington University
College of Saint Rose
CUNY City College
Eastern Washington University
Grand Valley State University
Olivet Nazarene University
CUNY-Queens College
Slippery Rock University
Sul Ross University
SUNY New Paltz
SUNY Oswego
University of Minnesota-Duluth
University of North Carolina at Pembroke
University of Northern Iowa

University of Wisconsin-Oshkosh
Weber State University
Wesleyan University
Western Kentucky University
Western Washington University

Exit Survey Departments (Master's/M):

California State University-Monterey Bay
Humboldt State University
Pacific Lutheran University
University of Tennessee at Martin

Exit Survey Departments (Master's/S):

Black Hills State University
Concord University
SUNY Geneseo
SUNY Oneonta
University of Hawaii-Hilo

Doctoral Universities-Highest Research Activity (DU/R1)

Doctoral Universities-Higher Research Activity (DU/R2)

Doctoral Universities-Moderate Research Activity (DU/R3)

Includes institutions that awarded at least 20 research/scholarship doctoral degrees during the update year (this does not include professional practice doctoral-level degrees, such as the JD, MD, PharmD, DPT, etc). Excludes Special Focus Institutions and Tribal Colleges.

Doctorate-granting institutions were assigned to one of three categories based on a measure of research activity. The "Shorthand" labels for the Doctoral Universities were restored in the 2015 update to numeric sequences to denote that each one is based on differences in quantitative levels. For more information about the methodology to determine the level of research activity, please visit the description of the Basic Classification Methodology (http://carnegieclassifications.iu.edu/methodology/basic.php).

Exit Survey Departments (DU/R1):

California Institute of Technology
Clemson University
Colorado State University
Cornell University
George Mason University
Georgia Institute of Technology
Harvard University

Iowa State University
Michigan State University
North Carolina State University
Northwestern University
Ohio State University
Pennsylvania State University
Purdue University
Rutgers University
Stanford University
Temple University
Texas Tech University
Tufts University
University of Arizona
University of California-Berkeley
University of California-Davis
University of California-Los Angeles
University of California-Riverside
University of California-San Diego
University of California-Santa Barbara
University of California-Santa Cruz
University of Colorado at Boulder
University of Connecticut
University of Delaware
University of Florida
University of Georgia
University of Hawaii-Manoa
University of Houston
University of Illinois
University of Kentucky
University of Maryland
University of Massachusetts
University of Michigan
University of Minnesota
University of Nebraska-Lincoln
University of Rochester
University of South Carolina
University of South Florida
University of Texas at Arlington
University of Texas at Austin
University of Texas at Dallas
University of Utah
University of Virginia
University of Washington
University of Wisconsin-Madison
Virginia Polytechnic Institute and State University
Yale University

Exit Survey Departments (DU/R2):

Baylor University
Bowling Green State University
College of William and Mary
Colorado School of Mines
East Carolina University
Miami University of Ohio
Michigan Technological University
Mississippi State University
Missouri University of Science and Technology
Montana State University

North Dakota State University
Northern Illinois University
Oklahoma State University
Old Dominion University
San Diego State University
Southern Methodist University
St. Louis University
University of Alabama
University of Alaska-Fairbanks
University of Colorado Denver
University of Louisiana at Lafayette
University of Memphis
University of Missouri
University of North Dakota
University of South Alabama
University of Vermont
University of Wyoming
Western Michigan University

Exit Survey Departments (DU/R3):

Boise State University
Indiana University of Pennsylvania
Middle Tennessee State University
Tennessee Tech University
University of Louisiana at Monroe
University of Nebraska-Omaha

Special Focus Institutions —
Schools of Engineering (Spec/Engg)

The special-focus designation was based on the concentration of degrees in a single field of set of related fields, at both the undergraduate and graduate levels. Institutions were determined to have a special focus with concentrations of at least 75% of undergraduate and graduate degrees. Excludes Tribal Colleges.

Exit Survey Departments (Spec/Engg)*:

South Dakota School of Mines and Technology

*Institutions in this classification where not included in comparisons using the Carnegie Classification system due to the small number of institutions in the Exit Survey belonging to the particular classification.